T0237448

Compact Textbooks in Mathematics

 Birkhäuser

Compact Textbooks in Mathematics

This textbook series presents concise introductions to current topics in mathematics and mainly addresses advanced undergraduates and master students. The concept is to offer small books covering subject matter equivalent to 2- or 3-hour lectures or seminars which are also suitable for self-study. The books provide students and teachers with new perspectives and novel approaches. They feature examples and exercises to illustrate key concepts and applications of the theoretical contents. The series also includes textbooks specifically speaking to the needs of students from other disciplines such as physics, computer science, engineering, life sciences, finance.

- **compact**: small books presenting the relevant knowledge
- **learning made easy**: examples and exercises illustrate the application of the contents
- **useful for lecturers**: each title can serve as basis and guideline for a semester course/lecture/seminar of 2–3 hours per week.

More information about this series at http://www.springer.com/series/11225

Christian Clason

Introduction to Functional Analysis

 Birkhäuser

Christian Clason
Fakultät für Mathematik
Universität Duisburg-Essen
Essen, Germany

ISSN 2296-4568 ISSN 2296-455X (electronic)
Compact Textbooks in Mathematics
ISBN 978-3-030-52783-9 ISBN 978-3-030-52784-6 (eBook)
https://doi.org/10.1007/978-3-030-52784-6

Mathematics Subject Classification: 46-01

This book is published under the imprint Birkhäuser, www.birkhauser-science.com, by the registered
company Springer Nature Switzerland AG.
The registered company address is: Gewerbestrasse 11, 6330 Cham, Switzerland

Preface

*Functional analysis is the continuation of linear algebra by
other means.*

The development of functional analysis in the early twentieth century was motivated
by the desire for general results on the solvability of differential equations. Instead
of solving concrete differential equations like $f'' + f = g$ for a given g by specific
techniques, people wanted to know which properties of a differential equation or
a right-hand side g were necessary for such a solution to exist. The crucial insight
here was to consider functions as points in a vector space, on which the mapping
$D : f \mapsto f'' + f$ defines a linear *differential operator*. A similar step from
concrete systems of linear equations to the abstract linear equation $Ax = b$ for
a matrix A and a vector b is the basis of linear algebra. The question was then
about the properties of D required for unique solvability of $Df = g$, in analogy
to the injectivity and surjectivity of A or the absence of 0 as an eigenvalue of A.
Here, the main difficulty lies in the fact that many of the fundamental results of
linear algebra are based on the finite-dimensionality of the involved vector spaces
(e.g., using the rank-nullity theorem). However, this is no longer the case for vector
spaces of functions, and it becomes necessary to consider these algebraic concepts
in combination with topological concepts such as convergence and compactness.
One of the central themes of this book is to work out which algebraic, metric,
topological, and geometric properties can serve as substitutes for the missing finite-
dimensionality, and what role precisely these play for the individual results. The fact
that this combination leads to extremely rich structures is what makes functional
analysis so appealing and has led to it becoming an essential foundation for modern
applied mathematics, from the theory and numerical solution of partial differential
equations, through optimization and probability theory, to medical imaging and
mathematical image processing.

The contents of this book correspond exactly to those of 26 lectures of 90 min
each (of which not too many should fall on a public holiday) in the fourth semester
of a bachelor's program in mathematics; it therefore cannot and is not supposed

to be a substitute for more comprehensive textbooks such as [7, 15, 22, 29–31] or the more recent [3, 23] (which are closer to the aims of this book). Its aim is rather to present a concise, streamlined, and rigorous development of the essential *structural* results that are important in particular throughout applied mathematics and thus build a solid foundation for mathematical lectures on the above-mentioned topics. All further results (e.g., on quotient spaces or Fredholm operators) are treated only insofar as they are needed or are useful for significantly simplifying a proof. After the tight corset of a lecture had been removed, many more beautiful results could have been included; this temptation was resisted. Readers are therefore certain to miss a favorite result; a particularly conspicuous gap is details on Lebesgue and Sobolev spaces, which are left to lectures on measure and integration theory (e.g., based on [5, Chapters XII and XIII]) and on partial differential equations, respectively. (A nice treatment can also be found in [3].)

The structure of the book as well is based on the desire to draw as clear and direct a line as possible to the main results. To this end, as far as possible, related topics are treated together, and more general results are exploited that are to be proved anyway. This leads to numerous dependencies but is likely in the spirit of functional analysis as an abstract structural theory (motivated by application in other branches of mathematics). Of course, some freedom remains even under these constraints that others might have used differently, in particular for an earlier treatment of Hilbert spaces. Specifically, it is possible to cover Chap. 15 directly after Chap. 3, with the exception of the Lax–Milgram theorem (Theorem 15.11, which requires Corollary 9.8) and the Fischer–Riesz theorem (Corollary 15.16, which requires the concept of isomorphism from Chap. 4). Chapter 16 requires the Hahn–Banach theorem (Theorem 8.1); the concept of a Hilbert-space adjoint operator is here defined via that of the (Banach space) adjoint operator but could also be introduced independently. Finally, Chap. 17 continues seamlessly from Chap. 14.

The chosen arrangement also corresponds to the mathematical tradition of presenting a theory without any reference to its historical development. This often allows a clearer view of the central structures and a concentration on concepts that have proven to be particularly fruitful. Conversely, one risks losing sight of the fact that these structures and concepts were developed by people who now only appear as eponyms for theorems. In particular, those are neglected who in initial phases produced groundbreaking results that were later subsumed in more general statements. This is especially true for functional analysis, which was brought to essentially the state as presented here by a comparatively small number of people within a few decades (between about 1900 and 1940). Rather than overloading this book with (even more) footnotes, the reader is referred to the literature on this point. An extensive account of the history of functional analysis can be found in [9], a short summary of which is contained (and more easily available) as an appendix in [2]. The development of the results treated in this book is described more accessibly— and more conversationally—in [12, Part XIX]. An even more extensive account up to the present is given in [18]; this work also contains a detailed chronology as well as numerous quotations from and references to original works that are often still worth reading. Historical and biographical remarks can also be found in the appendices to the individual chapters of [28] (which are also worth reading).

This book is based mainly on [1, 4, 14, 27, 28], which are also recommended for supplemental reading. Thanks are due to Martin Brokate and Gerd Wachsmuth for their lecture notes and for helpful comments, Otmar Scherzer and Anton Schiela for helpful comments as well, and Remo Kretschmann for help with the exercises. Finally, David P. Kramer's expert copy-editing was much appreciated.

Essen, Germany   Christian Clason
May 2020

Contents

Part I   Topological Basics

1   Metric Spaces 3

2   Compact Sets 9

Part II   Linear Operators Between Normed Spaces

3   Normed Vector Spaces 19

4   Linear Operators 35

5   The Uniform Boundedness Principle 45

6   Quotient Spaces 55

Part III   Dual Spaces and Weak Convergence

7   Linear Functionals and Dual Spaces 63

8   The Hahn–Banach Theorem 71

9   Adjoint Operators 83

10   Reflexivity 93

11   Weak Convergence 99

Part IV   Compact Operators Between Banach Spaces

12   Compact Operators 111

13   The Fredholm Alternative 119

14   The Spectrum 123

Part V   Hilbert Spaces

15   Inner Products and Orthogonality 135

16   The Riesz Representation Theorem 149

17   Spectral Decomposition in Hilbert Spaces 157

References 165

Index 167

Part I

Topological Basics

Metric Spaces

1

We begin by collecting the fundamental topological structures needed for functional analysis. The core concepts are assumed to be familiar from a course in real analysis; a detailed treatment including proofs can be found in textbooks such as [8, Chapter 1] and [20, Chapter 2].

Definition 1.1

Let X be a set. A *metric* on X is a mapping $d : X \times X \to \mathbb{R}$ that for all $x, y, z \in X$ satisfies

  (i) $d(x, y) = 0$ if and only if $x = y$ *(nondegeneracy)*;
  (ii) $d(x, y) = d(y, x)$ *(symmetry)*;
 (iii) $d(x, z) \leq d(x, y) + d(y, z)$ *(triangle inequality)*.

In this case, the pair (X, d) is called a *metric space*. If the metric is clear from the context, the metric space is denoted by just X.

A metric is the mathematical formalization of the intuitive notion of distance. (Note that we have not yet introduced a vector space structure, which allows taking the difference of two elements.) The properties directly imply that a metric is always nonnegative: for all $x, y \in X$, we have that

$$2d(x, y) = d(x, y) + d(x, y) = d(x, y) + d(y, x) \geq d(x, x) = 0.$$

© Springer Nature Switzerland AG 2020
C. Clason, *Introduction to Functional Analysis*, Compact Textbooks
in Mathematics, https://doi.org/10.1007/978-3-030-52784-6_1

Example 1.2 The following are canonical examples of metric spaces:

(i) the *Euclidean metric*: $X = \mathbb{R}^n$ or $X = \mathbb{C}^n$ and

$$d(x, y) := \left(\sum_{i=1}^{n} |x_i - y_i|^2 \right)^{\frac{1}{2}} ;$$

(ii) the *relative metric*: if (X, d) is a metric space and $A \subset X$, then $(A, d|_{A \times A})$ is also a metric space, where $d|_{A \times A}$ denotes the restriction of d to $A \times A$;

(iii) the *product metric*: if (X, d_X) and (Y, d_Y) are metric spaces, then $(X \times Y, d_{X \times Y})$ is also a metric space for

$$d_{X \times Y}((x_1, y_1), (x_2, y_2)) := d_X(x_1, x_2) + d_Y(y_1, y_2),$$

as well as for

$$d_{X \times Y}((x_1, y_1), (x_2, y_2)) := \max\{d_X(x_1, x_2), d_Y(y_1, y_2)\};$$

(iv) the *discrete metric*: X is an arbitrary set and

$$d(x, y) := \begin{cases} 0 & \text{if } x = y, \\ 1 & \text{if } x \neq y. \end{cases}$$

In the following, let (X, d) always be a metric space. We now define for $x \in X$ and $r > 0$

(i) the *closed ball* $B_r(x) := \{y \in X : d(x, y) \leq r\}$,
(ii) the *open ball* $U_r(x) := \{y \in X : d(x, y) < r\}$,

around x with radius r. With the help of these balls, we can define the following fundamental topological concepts.

Definition 1.3

A set $U \subset X$ is called

(i) *open* if for all $x \in U$ there exists an $\varepsilon > 0$ with $U_\varepsilon(x) \subset U$;
(ii) a *neighborhood* of $x \in U$ if there exists an open set O with $x \in O \subset U$;
(iii) a *neighborhood* of $A \subset U$ if U is a neighborhood of all $x \in A$.

A set $C \subset X$ is called *closed* if the complement $X \setminus C$ is open.

It follows from the definition that open balls are open and closed balls are closed. Furthermore, the intersection of finitely many open sets as well as the union of arbitrarily (even infinitely) many open sets is open. Two metric spaces (X, d_1)

and (X, d_2) are called *equivalent* if they contain the same open sets (e.g., the two definitions in Example 1.2 (iii)). The set of all open subsets is called a *topology* on X.

Clearly, X as well as the empty set $\emptyset$ are both open and closed; other sets may be neither open nor closed. In this case, we can construct from them other sets that are open or closed.

Definition 1.4

For $A \subset X$, we define

(i) the *interior*   int $A := \bigcup_{\{U \subset A : U \text{ open}\}} U$;

(ii) the *closure*   cl $A := \bigcap_{\{C \supset A : C \text{ closed}\}} C$.

In the literature, one can also often find the notation $A^o := \text{int } A$ and $\overline{A} := \text{cl } A$.

If $x \in \text{int } A$, then x is called an *interior point* of A; in other words, $x \in A$ is an interior point if there exists an $\varepsilon > 0$ with $U_\varepsilon(x) \subset A$. If cl $A = X$, then A is called *dense* in X. If there exists a countable dense $A \subset X$, then X is called *separable*.

The definition implies that the interior of A is always open and the closure of A always closed. In particular, A is open if and only if $A = \text{int } A$, and closed if and only if $A = \text{cl } A$.

Finally, we call $A \subset X$ *bounded* if the *diameter*

$$\text{diam}(A) := \sup_{x,y \in A} d(x, y)$$

is finite.

A metric allows defining a notion of convergence for sequences.

Definition 1.5

A sequence $\{x_n\}_{n \in \mathbb{N}} \subset X$ *converges in X* to a *limit* $x \in X$ if one of the following equivalent properties holds:

(i) for all $\varepsilon > 0$ there exists an $N \in \mathbb{N}$ with $d(x_n, x) \leq \varepsilon$ for all $n \geq N$;

(ii) for every neighborhood U of x there exists an $N \in \mathbb{N}$ with $x_n \in U$ for all $n \geq N$.

In this case we write $x_n \to_{(X,d)} x$, or simply $x_n \to x$ if the metric space is obvious from the context.

The equivalence of these definitions is a straightforward consequence of the definition of neighborhoods. The definition also implies the uniqueness of the limit. We further have for $A \subset X$ that

$$\text{cl } A = \{x \in X : \text{there exists } \{x_n\}_{n \in \mathbb{N}} \text{ with } x_n \to x\}. \tag{1.1}$$

In particular, A is closed if and only if the limit of every convergent sequence $\{x_n\}_{n\in\mathbb{N}} \subset X$ is an element of A. Furthermore, two metric spaces (X, d_1) and (X, d_2) are equivalent if and only if they have the same convergent sequences (with identical respective limits).

Definition 1.6

Let $\{x_n\}_{n\in\mathbb{N}} \subset X$ be a sequence.

(i) If $\{n_k\}_{k\in\mathbb{N}} \subset \mathbb{N}$ is a strictly increasing sequence, then the sequence $\{x_{n_k}\}_{k\in\mathbb{N}}$ is called a *subsequence* of $\{x_n\}_{n\in\mathbb{N}}$.
(ii) If $\{x_n\}_{n\in\mathbb{N}}$ contains a convergent subsequence with limit $x \in X$, then x is called a *cluster point* of $\{x_n\}_{n\in\mathbb{N}}$.
(iii) If for all $\varepsilon > 0$ there exists an $N \in \mathbb{N}$ with $d(x_m, x_n) \leq \varepsilon$ for all $m, n \geq N$, then $\{x_n\}_{n\in\mathbb{N}}$ is called a *Cauchy sequence*.

Exactly as for real sequences, one can show that every Cauchy sequence has at most one cluster point. Hence every convergent sequence is a Cauchy sequence, but the converse is not true in general. Metric spaces in which every Cauchy sequence is convergent are called *complete*; as we will see in Part II, this is a fundamental property with far-reaching consequences. For example, both $\mathbb{R}^n$ and $\mathbb{C}^n$, either endowed with the Euclidean or the product metrics from Example 1.2 (iii), are complete. Furthermore, closed subsets of complete metric spaces, endowed with the relative metric, are again complete metric spaces. Note that equivalent metric spaces have the same convergent sequences but need not have the same Cauchy sequences—equivalence therefore does not conserve completeness. (Hence equivalence is a *metric* and not a *topological* property.)

Similar to the convergence of sequences, one can extend the continuity of mappings to metric spaces.

Definition 1.7

Let (X, d_X) and (Y, d_Y) be metric spaces. A mapping $f : X \to Y$ is called *continuous* at $x \in X$ if one of the following equivalent properties holds:

(i) for all $\varepsilon > 0$ there exists a $\delta > 0$ with $f(B_\delta(x)) \subset B_\varepsilon(f(x))$ (equivalently: $f(U_\delta(x)) \subset U_\varepsilon(f(x))$);
(ii) for every neighborhood V of $f(x)$ there exists a neighborhood U of x with $f(U) \subset V$;
(iii) for every sequence $\{x_n\}_{n\in\mathbb{N}} \subset X$ with $x_n \to x$ in X one has $f(x_n) \to f(x)$ in Y.

We call f continuous on X if f is continuous at x for all $x \in X$, and *bounded* if $\sup_{x,y\in X} d_Y(f(x), f(y)) < \infty$.

This definition formalizes the intuitive notion that a continuous function f maps points near x to points near $f(x)$. Again, the equivalence of these definitions follows from those of neighborhoods and convergence in metric spaces. An alternative characterization that will later be useful is the following.

Theorem 1.8
Let (X, d_X) and (Y, d_Y) be metric spaces. Then a mapping $f : X \to Y$ is continuous if and only if every open set $V \subset Y$ has an open preimage

$$f^{-1}(V) := \{x \in X : f(x) \in V\}.$$

However, images of open sets need *not* be open. Taking complements, we obtain

---

Corollary 1.9
A mapping $f : X \to Y$ is continuous if and only every closed set $V \subset Y$ has a closed preimage $f^{-1}(V)$.

If X is a metric space, then the space of all continuous, bounded, real-valued (or complex-valued) functions,

$$C_b(X) := \{f : X \to \mathbb{R} : f \text{ continuous and bounded}\},$$

together with

$$d(f, g) := \sup_{x \in X} |f(x) - g(x)|, \tag{1.2}$$

is a complete metric space. In this case, Definition 1.5 corresponds to the uniform convergence of sequences of functions.

Metric spaces are not the most general setting in which the above concepts can be introduced. Instead of defining open sets via open balls (i.e., via a metric), one can define them axiomatically: define a *topology* τ on X as a system of subsets of X that contains X and $\emptyset$ and is closed under unions and finite intersections; the pair $(X; \tau)$ is then called a *topological space*. Similarly, one defines convergence of sequences and continuity of mappings in topological spaces directly via Definition 1.5 (ii) and Definition 1.7 (ii), respectively; however, in this case Definition 1.7 (ii) and (iii) are no longer necessarily equivalent (the latter is then called *sequential continuity*). Details can be found in, e.g., [8]. Topological spaces are relevant, for example, in studying the pointwise (but not uniform) convergence of sequences of functions, which in general cannot be expressed in terms of a metric (i.e., such spaces are not *metrizable*).

Problems

Problem 1.1 *(Balls and neighborhoods)*

(i) Let (X, d) be a metric space and $x \in X$. Show that every neighborhood of x contains a closed ball $B_r(x)$ around x with radius $r > 0$.
(ii) Give an example for a metric space (X, d) for which there exist an $x \in X$ and an $r > 0$ such that $\operatorname{cl} U_r(x) \neq B_r(x)$.

Problem 1.2 *(Stereographic projection)*
Let d_1 be the Euclidean metric on $\mathbb{R}$ and let $d_2 \colon \mathbb{R} \times \mathbb{R} \to [0, \infty)$ be defined as

$$d_2(s, t) = |\arctan t - \arctan s| \quad \text{for all } s, t \in \mathbb{R}.$$

Show that

(i) d_2 is a metric on $\mathbb{R}$;
(ii) the metric spaces $(\mathbb{R}, d_1)$ and $(\mathbb{R}, d_2)$ are equivalent;
(iii) the metric space $(\mathbb{R}, d_2)$ is not complete.

Problem 1.3 *(Interior of intersections)*
Let (X, d) be a metric space and $A_1, \ldots, A_n \subset X$. Show that

$$\operatorname{int}\left(\bigcap_{k=1}^{n} A_k\right) = \bigcap_{k=1}^{n} \operatorname{int} A_k.$$

Does this also hold for the intersection of infinitely many subsets?

Problem 1.4 *(Separable subsets)*
Show that every subset A of a separable metric space M is separable as well.

Problem 1.5 *(Discrete balls)*
Let (X, d) be a discrete metric space (i.e., endowed with the discrete metric). Characterize

(i) all open balls;
(ii) all dense subsets.

Problem 1.6 *(A convergence principle[1])*
Let (X, d) be a metric space. Show that a sequence $\{x_n\}_{n \in \mathbb{N}}$ in X converges to some $x \in X$ if and only if every subsequence $\{x_{n_k}\}_{k \in \mathbb{N}}$ of $\{x_n\}_{n \in \mathbb{N}}$ itself contains a subsequence that converges to x.

---

[1]This ubiquitous convergence principle is known as a *subsequence–subsequence argument*. Note that to conclude convergence of the full sequence, all convergent subsequences must have the *same* limit.

Compact Sets

2

A fundamental metric property is compactness; informally, continuous functions on compact sets behave almost as nicely as functions on finite sets.

Throughout the following, let (X, d) be again a metric space. We first define several related notions of compactness.

Definition 2.1

A set $K \subset X$ is called

(i) *(open cover)* compact if every open cover of K contains a finite subcover, i.e., if for every family $\{U_i\}_{i \in I}$ of sets with $U_i \subset X$ open and $K \subset \bigcup_{i \in I} U_i$ there exists a *finite* subset $J \subset I$ with $K \subset \bigcup_{j \in J} U_j$;

(ii) *sequentially compact* if every sequence $\{x_n\}_{n \in \mathbb{N}} \subset K$ contains a convergent subsequence $\{x_{n_k}\}_{k \in \mathbb{N}}$ with limit $x \in K$;

(iii) *precompact* (or *totally bounded*) if for all $\varepsilon > 0$ there exists a finite cover of open balls with radius ε, i.e., there exist $N \in \mathbb{N}$ and $x_1, \ldots, x_N \in K$ such that $K \subset \bigcup_{n=1}^{N} U_\varepsilon(x_i)$.

A metric space (K, d) where K is compact is also called a *compact space*.

Definition (i) is technical but indicates why compactness is a useful substitute for finiteness: to verify a property, it suffices to consider *finitely many* open neighborhoods. Definition (ii), on the other hand, is the most useful property in practice, since it allows one to extract a cluster point from *any* sequence.[1] Note that definition (iii) implies that subsets of precompact sets are again precompact (which

---

[1] A nice presentation of the historical development of the notion of compactness can be found in [19].

© Springer Nature Switzerland AG 2020
C. Clason, *Introduction to Functional Analysis*, Compact Textbooks
in Mathematics, https://doi.org/10.1007/978-3-030-52784-6_2

need not be the case for compact sets according to definition (i)). Furthermore, precompact sets are always bounded (as suggested by the alternative name).

In metric spaces, all three properties are equivalent.[2]

Theorem 2.2

For every $K \subset X$, the following properties are equivalent:

  (i) *K is compact;*
 (ii) *K is sequentially compact;*
(iii) *K is complete (with respect to the relative metric) and precompact.*

Proof. (i) $\Rightarrow$ (ii): We argue by contradiction. Assume that there exists a sequence $\{x_n\}_{n\in\mathbb{N}} \subset K$ without a cluster point. This means that for all $x \in K$ there exists an $r_x > 0$ such that $U_{r_x}(x)$ contains only finitely many elements x_n (otherwise we could extract a subsequence that converges by Definition 1.5 (ii), in contradiction to the assumption). Then the family $\{U_{r_x}(x)\}_{x\in K}$ forms an open cover of K, whose compactness yields the existence of a finite subcover $\{U_{r_{\tilde{x}_i}}(\tilde{x}_i)\}_{i=1,\dots,N}$ of K. Since each of these sets contains only finitely many elements of the sequence, this is also true for their (finite) intersection. Hence K itself can contain only finitely many elements of the sequence, in contradiction to $\{x_n\}_{n\in\mathbb{N}} \subset K$.

(ii) $\Rightarrow$ (iii): Since every Cauchy sequence with a cluster point is convergent and, by assumption, every sequence has a cluster point, (K, d) is complete by definition. Assume now that K is not precompact. Then there exists an $\varepsilon > 0$ such that K cannot be covered by finitely many balls of radius ε. Hence we can construct a sequence inductively by choosing $x_1 \in K$ arbitrary and

$$x_{n+1} \in K \setminus \bigcup_{i=1}^{n} U_\varepsilon(x_i), \qquad n \in \mathbb{N}.$$

(This construction is possible since by assumption, the set on the right-hand side is never empty.) This implies that every ball of radius ε contains at most one element of the sequence $\{x_n\}_{n\in\mathbb{N}}$, which therefore cannot have a cluster point. Thus K is not sequentially compact.

(iii) $\Rightarrow$ (i): This is the most difficult part. We again argue by contradiction. Let $\{U_i\}_{i\in I}$ be an open cover of K. We then define the system $\mathcal{B}$ of all sets that can be covered only by infinitely many U_i and show that the assumption $K \in \mathcal{B}$ leads to a contradiction. To this end, we construct by induction open balls $B_n := U_{2^{-n}}(x_n)$ with $B_n \cap B_{n-1} \neq \emptyset$ and $B_n \in \mathcal{B}$. For $n = 1$, we use the precompactness of K to choose finitely many balls with radius $\varepsilon = \frac{1}{2}$ whose union covers K. By assumption, there has to be at least one ball in $\mathcal{B}$ among them (otherwise $K \notin \mathcal{B}$); we denote it by $B_1 = U_{\frac{1}{2}}(x_1)$. Let now $B_{n-1} \in \mathcal{B}$ be chosen as desired. Then there in turn exist finitely many open balls with radius 2^{-n} whose union covers K and

---

[2]This is no longer necessarily the case in topological spaces; see, e.g., [25, Examples 42 and 105].

has nonempty intersection with B_{n-1}. Since $B_{n-1} \in \mathcal{B}$, at least one of these balls has to be in $\mathcal{B}$, which we denote by $B_n = U_{2^{-n}}(x_n)$. This defines a Cauchy sequence $\{x_n\}_{n \in \mathbb{N}} \subset K$, since we have for all $n \in \mathbb{N}$ that

$$d(x_n, x_{n+1}) < 2^{-n} + 2^{-(n+1)} < 2^{-n+1}$$

and hence

$$d(x_n, x_m) < 2^{-n+1} \qquad \text{for all } n, m \text{ with } m \geq n.$$

Since (K, d) is complete, $x_n \to x \in K$. The covering property then implies that $x \in U_j$ for some $j \in I$. Since U_j is open, there exists an $\varepsilon > 0$ with $x \in U_\varepsilon(x) \subset U_j$; it follows for all $n \in \mathbb{N}$ sufficiently large that

$$B_n = U_{2^{-n}}(x_n) \subset U_\varepsilon(x) \subset U_i,$$

in contradiction to $B_n \in \mathcal{B}$. □

In particular, the proof of (iii) ⇒ (i) shows that every precompact set contains a Cauchy sequence.

Property (iii) can be interpreted as a stronger version of closedness plus boundedness. In fact, both properties hold for compact sets.

Corollary 2.3
If $K \subset X$ is compact, then K is bounded and closed.

Proof. By Theorem 2.2 (iii), K is precompact and hence bounded (since $\text{diam}(K) < N\varepsilon$). Theorem 2.2 (ii) further yields that every sequence has a cluster point contained in K; hence the limit of every convergent sequence (which is then the only cluster point) is in K, which is therefore closed. □

For $X = \mathbb{R}^n$, the converse also holds. We will show this using the following useful lemma.

Lemma 2.4
If $K \subset X$ is compact and $C \subset K$ is closed, then C is compact as well.

Proof. Let $\{U_i\}_{i \in I}$ be an open cover of C. Since C is closed, $X \setminus C$ is open, and hence $\{U_i\}_{i \in I} \cup (X \setminus C)$ is an open cover of K. Since K is compact, there exists a finite subcover $\{U_n\}_{n=1,\dots,N} \cup (X \setminus C)$ of K and thus of $C \subset K$ as well. But $C \cap (X \setminus C) = \emptyset$, so that $X \setminus C$ can obviously not cover C. Thus $\{U_n\}_{n=1,\dots,N}$ has to be a finite subcover of C, implying that C is compact. $\square$

To obtain the desired result, we have to endow $\mathbb{R}^n$ with one of the standard metrics such as Example 1.2 (i).

Theorem 2.5 (Heine–Borel)
A subset of $\mathbb{R}^n$ is compact with respect to the Euclidean metric if and only if it is closed and bounded.

Proof. Due to Theorem 2.3, we have only to show that closed and bounded subsets of $\mathbb{R}^n$ are compact. Let therefore $C \subset \mathbb{R}^n$ be closed and bounded. We will use the following results from real analysis: A set is bounded with respect to the Euclidean metric if and only if its elements are bounded componentwise, i.e., if there exists an $M > 0$ such that

$$C \subset \prod_{i=1}^{n} [-M, M] =: K.$$

Furthermore, the closed interval $[-M, M]$ is compact. We now show via a diagonal sequence argument that this implies the (sequential) compactness of the n-dimensional unit cube K. This calls for some special notation. Let $\{x_k\}_{k \in \mathbb{N}} \subset K$ be a sequence, where we write $x_k = (x_k^1, \dots, x_k^n)$ for $x_k \in \mathbb{R}^n$. Then $\{x_k^1\}_{k \in \mathbb{N}} \subset [-M, M]$, and hence there exists a convergent subsequence, which we denote by $\{x_k^1\}_{k \in \mathbb{N}_1}$ for an infinite subset $\mathbb{N}_1 \subset \mathbb{N}$, with limit $x^1 \in [-M, M]$. We now consider the sequence $\{x_k^2\}_{k \in \mathbb{N}_1} \subset [-M, M]$, which in turn contains a subsequence $\{x_k^2\}_{k \in \mathbb{N}_2}$ for $\mathbb{N}_2 \subset \mathbb{N}_1$ with limit $x^2 \in [-M, M]$. Proceeding along these lines, we finally obtain a set $\mathbb{N}_n \subset \cdots \subset \mathbb{N}_1 \subset \mathbb{N}$ such that $\{x_k^n\}_{k \in \mathbb{N}_n}$ converges to some $x^n \in [-M, M]$. Since every subsequence of a convergent sequence converges to the same limit, this yields a subsequence $\{x_k\}_{k \in \mathbb{N}_n}$ with limit $x := (x^1, \dots, x^n) \in \prod_{i=1}^{n}[-M, M]$, where the convergence is componentwise. However, since sequences in $\mathbb{R}^n$ converge with respect to the Euclidean metric if and only if they converge componentwise, this implies the sequential compactness of K. The compactness of C then follows from Lemma 2.4. $\square$

The Heine–Borel theorem therefore crucially relies on the equivalence of metric and componentwise convergence (and boundedness) and thus no longer holds in infinite-dimensional metric spaces. (We will later see a counterexample.) This reflects one of the fundamental complications in functional analysis compared to linear algebra.[3]

One often formulates Theorem 2.5 directly in terms of sequential compactness.

Corollary 2.6 (Bolzano–Weierstraß theorem)
Every bounded sequence in $\mathbb{R}^n$ contains a convergent subsequence.

If the set K in Theorem 2.2 (iii) is not closed, we still have the following weaker result.

Theorem 2.7
Let (X, d) be complete and $A \subset X$. Then the following properties are equivalent:

  (i) *A is precompact;*
 (ii) *A is relatively compact, i.e., cl A is compact;*
(iii) *every sequence in A contains a convergent subsequence (whose limit need not lie in A).*

Proof. (iii) $\Rightarrow$ (ii): Let $\{x_n\}_{n\in\mathbb{N}}$ be a sequence in cl A. Then (1.1) implies that for every x_n, there exists a sequence $\{x_{n,k}\}_{k\in\mathbb{N}} \subset A$ with $x_{n,k} \to x_n$ as $k \to \infty$; in other words, for every $\varepsilon > 0$ there exist $N_n \in \mathbb{N}$ and $\tilde{x}_n := x_{n,N}$ with $d(x_n, \tilde{x}_n) \le \varepsilon/2$. Consider now the sequence $\{\tilde{x}_n\}_{n\in\mathbb{N}} \subset A$, which by assumption contains a convergent subsequence $\{\tilde{x}_{n_k}\}_{k\in\mathbb{N}} \subset A$ with limit $x \in$ cl A (since the limit of every convergent sequence in A is by definition an element of cl A). Hence for every $\varepsilon > 0$ there exists an $N \in \mathbb{N}$ with $d(x, \tilde{x}_{n_k}) \le \varepsilon/2$ for all $k \ge N$, which implies that

$$d(x, x_{n_k}) \le d(x, \tilde{x}_{n_k}) + d(\tilde{x}_{n_k}, x_{n_k}) \le \varepsilon \qquad \text{for all } k \ge N,$$

i.e., the subsequence $\{x_{n_k}\}_{k\in\mathbb{N}}$ converges to $x \in$ cl A, and therefore cl A is compact.

(ii) $\Rightarrow$ (i): If cl A is compact, then cl A is precompact by Theorem 2.2. Since subsets of precompact sets are again precompact, $A \subset$ cl A is precompact as well.

(i) $\Rightarrow$ (iii): Let $\{x_n\}_{n\in\mathbb{N}} \subset A$ be a sequence. If A is precompact, then so is the subset of all elements of this sequence. As in the proof of Theorem 2.2, this implies that $\{x_n\}_{n\in\mathbb{N}}$ contains a Cauchy subsequence, which converges since (X, d) is complete. □

We now consider continuous functions on compact sets.

---

[3]Even on $\mathbb{R}^n$, it is possible to construct metrics for which one of these equivalences and hence the claim of the theorem is violated.

Theorem 2.8
*Let (X, d_X) and (Y, d_Y) be metric spaces and let $f : X \to Y$ be continuous. If $K \subset X$
is compact, then $f(K) \subset Y$ is compact.*

Proof. Let $\{U_i\}_{i \in I}$ be an open cover of $f(K)$. Since f is continuous, the preimages
$\{f^{-1}(U_i)\}_{i \in I}$ are open as well and form a cover of K (otherwise there would exist an $x \in K$
with $x \notin f(K)$, which is not possible by definition of the preimage). The compactness of K
then implies the existence of a finite subcover $\{f^{-1}(U_i)\}_{i \in J}$, so that $\{U_i\}_{i \in J}$ is the desired
finite subcover of $f(K)$.   □

A consequence of particular relevance for optimization is the fact that continuous
real-valued functions on compact sets always attain a maximum and a minimum.

Corollary 2.9 (Weierstraß theorem)
*Let (K, d) be a compact space and let $f : K \to \mathbb{R}$ be continuous. Then there exist
$a, b \in K$ such that $f(a) \le f(x) \le f(b)$ for all $x \in K$.*

Proof. By Theorem 2.8, $f(K) \subset \mathbb{R}$ is compact and therefore bounded and closed. In
particular, the boundedness implies that $\alpha := \inf f(K)$ and $\beta := \sup f(K)$ are finite. The
properties of infima and suprema then yield the existence of sequences $\{x_n\}_{n \in \mathbb{N}} \subset K$ with
$f(x_n) \to \alpha$ and $\{y_n\}_{n \in \mathbb{N}} \subset K$ with $f(y_n) \to \beta$. The closedness of $f(K)$ now implies that
$\alpha, \beta \in f(K)$ and hence the claim.   □

In particular, continuous functions on compact sets are always bounded; therefore
in compact spaces (K, d), we have

$$C_b(K) = C(K) := \{f : K \to \mathbb{R} : f \text{ is continuous}\} .$$

It is not yet clear whether infinite-dimensional spaces in fact can contain compact
subsets. We will show this specifically for $C(K)$, for which we need the following
lemma.

Lemma 2.10
Every compact space is separable.

Proof. Let (K, d) be compact. We have to show that there exists a countable dense subset. We construct this as follows. By Theorem 2.2, K is precompact, i.e., for every $\varepsilon > 0$, there exists a finite cover with balls of radius ε. Denote for given $n \in \mathbb{N}$ and $\varepsilon := \frac{1}{n}$ the set of centers of these balls by P_n. Since all P_n are finite, the set $P := \bigcup_{n \in \mathbb{N}} P_n$ is countable. Let now $x \in K$ be arbitrary. For every $\varepsilon > 0$ we can then choose an $n \in \mathbb{N}$ with $\frac{1}{n} < \varepsilon$, and the covering property yields the existence of an $x_n \in P_n \subset P$ with $d(x_n, x) < \varepsilon$. By definition, we then have $x_n \to x$, i.e., $x \in \text{cl } P$ and hence $K = \text{cl } P$. □

We can now give a characterization of precompactness in $C(K)$.

Theorem 2.11 (Arzelà–Ascoli)

Let (K, d) be a compact space and $A \subset C(K)$. If A is

(i) *bounded pointwise, i.e., the set $\{f(x) : x \in K\} \subset \mathbb{R}$ is bounded for all $f \in A$,*
(ii) *equicontinuous, i.e., for every $\varepsilon > 0$ there exists a $\delta > 0$ such that for all $f \in A$ and $x, y \in K$, $d(x, y) \leq \delta$ implies that $|f(x) - f(y)| \leq \varepsilon$,*

then A is precompact.

Proof. We use Theorem 2.7 (iii) and construct for a given sequence $\{f_n\}_{n \in \mathbb{N}} \subset A$ a convergent subsequence via a diagonal sequence argument. First, by Lemma 2.10 there exists a countable dense subset $\{x_1, x_2, \ldots\} =: X \subset K$. We set $f_n^0 := f_n$ and consider the sequence $\{f_n^0(x_1)\}_{n \in \mathbb{N}} \subset \mathbb{R}$. This sequence is bounded by assumption (i) and therefore contains a convergent subsequence by the Bolzano–Weierstraß theorem (Corollary 2.6), which we denote by $\{f_n^1(x_1)\}_{n \in \mathbb{N}}$. Proceeding along these lines we can thus find for every $j \in \mathbb{N}$ a subsequence $\{f_n^j\}_{n \in \mathbb{N}}$ such that $\{f_n^j(x_k)\}_{n \in \mathbb{N}}$ converges for all $k \leq j$. From this sequence of subsequences, we now construct a diagonal sequence by choosing $f_n^* := f_n^n$. This is a subsequence of $\{f_n\}_{n \in \mathbb{N}}$ as well as of $\{f_n^j\}_{n \in \mathbb{N}}$ for all $n \geq j$. Hence $\{f_n^*(x_j)\}_{n \in \mathbb{N}}$ converges for all $j \in \mathbb{N}$, i.e., pointwise on a dense subset.

We now use the equicontinuity to show that this convergence is even uniform. Since $C(K)$ is complete, it suffices to show that $\{f_n^*\}_{n \in \mathbb{N}}$ is a Cauchy sequence. To this end, let $\varepsilon > 0$ be arbitrary and choose $\delta > 0$ according to the definition of equicontinuity. Now K is precompact, and hence there exists a cover of K with finitely many open balls $U_1, \ldots, U_p$ with radius $\frac{\delta}{2}$. Since the set X is dense in K, each of these balls has to contain at least one element of X; to keep the notation concise, we will assume that $x_i \in U_i$ for all $i = 1, \ldots, p$. The convergence of the $\{f_n^*(x_i)\}_{n \in \mathbb{N}}$ then implies the existence of an $N \in \mathbb{N}$ with

$$|f_n^*(x_i) - f_m^*(x_i)| \leq \varepsilon \qquad \text{for all } m, n \geq N \text{ and } i = 1, \ldots, p.$$

Let now $x \in K$ be arbitrary. Then there exists a $j \in \{1, \ldots, p\}$ with $x \in U_j$, implying that $d(x, x_j) < \delta$ and hence by the equicontinuity of $\{f_n^*\}_{n \in \mathbb{N}} \subset A$ that

$$|f_n^*(x_j) - f_n^*(x)| \leq \varepsilon \qquad \text{for all } n \in \mathbb{N}.$$

Taking these together, we obtain for all $m, n \geq N$ that

$$|f_n^*(x) - f_m^*(x)| \leq |f_n^*(x) - f_n^*(x_j)| + |f_n^*(x_j) - f_m^*(x_j)| + |f_m^*(x_j) - f_m^*(x)| \leq 3\varepsilon.$$

Taking the supremum over all $x \in K$ now implies that $d(f_n^*, f_m^*) \leq 3\varepsilon$ for all $n, m \geq N$, i.e., that the subsequence $\{f_n^*(x_i)\}_{n \in \mathbb{N}}$ is a Cauchy sequence and hence convergent.   □

In fact, the converse holds as well; see, e.g., [7, Theorem 3.8].

Problems

Problem 2.1 *(Union of compact sets)*
Show that the union of finitely many compact sets is again compact. Give an example showing that this is no longer the case for infinitely many sets.

Problem 2.2 *(Closure of totally bounded sets)*
Show that the closure of a totally bounded set is again totally bounded.

Problem 2.3 *(Discrete compact sets)*
Show that a discrete metric space (X, d) is compact if and only if X is finite.

Problem 2.4 *(Compact sequences)*
Let (X, d) be a metric space and $\{x_n\}_{n \in \mathbb{N}} \subset X$ a convergent sequence with limit $x \in X$. Show that $\{x_n\}_{n \in \mathbb{N}} \cup \{x\}$ is compact.

Problem 2.5 *(Noncompact sets)*
Let $C([0, \pi])$ be the set of all continuous real-valued functions on the interval $[0, \pi]$ endowed with the *supremum metric*

$$d(f, g) = \sup_{x \in [0,\pi]} |f(x) - g(x)| \qquad \text{for all } f, g \in C([0, \pi]).$$

Show that the closed unit ball

$$B_C := \{f \in C([0, \pi]) : d(f, 0) \leq 1\}$$

is not (sequentially) compact.

Part II
Linear Operators Between Normed Spaces

Normed Vector Spaces

3

We now combine the topological and metric properties introduced in Part I with the algebraic structure of a vector space. As we will see in the following chapters, the property of completeness in particular will have far-reaching consequences.

We recall that a vector space X over a field $\mathbb{F}$ is a nonempty set that is closed with respect to addition of elements of X (called *vectors*) and multiplication by elements of $\mathbb{F}$ (called *scalars*) and that satisfies associative and distributive laws. In this book, we will restrict ourselves to the cases $\mathbb{F} = \mathbb{R}$ and $\mathbb{F} = \mathbb{C}$ (i.e., *real* and *complex vector spaces*).

Definition 3.1

Let X be a vector space over $\mathbb{F}$. A *norm* on X is a mapping $\|\cdot\|_X : X \to \mathbb{R}^+ := [0, \infty)$ that satisfies for all $x, y \in X$ and $\lambda \in \mathbb{F}$ the following properties:

  (i) $\|x\|_X = 0$ if and only if $x = 0 \in X$ *(nondegeneracy)*;
  (ii) $\|\lambda x\|_X = |\lambda|\|x\|_X$ *(homogeneity)*;
  (iii) $\|x + y\|_X \le \|x\|_X + \|y\|_X$ *(triangle inequality)*.

In this case, the pair $(X, \|\cdot\|_X)$ is called a *normed vector space*. If the norm is obvious from the context, we simply write X for the normed vector space. Conversely, if the space is obvious from the context, we simply write $\|\cdot\|$ for the norm.

Two norms $\|\cdot\|_1$ and $\|\cdot\|_2$ on X are called *equivalent* if there exist constants $c, C > 0$ such that

$$c\|x\|_1 \le \|x\|_2 \le C\|x\|_1 \qquad \text{for all } x \in X. \tag{3.1}$$

© Springer Nature Switzerland AG 2020
C. Clason, *Introduction to Functional Analysis*, Compact Textbooks
in Mathematics, https://doi.org/10.1007/978-3-030-52784-6_3

Before looking at examples, we will collect some fundamental properties. A norm on X induces a metric via

$$d(x, y) := \|x - y\| \qquad \text{for all } x, y \in X.$$

In this way, every normed vector space $(X, \|\cdot\|)$ corresponds in a canonical way to a metric space (X, d), which we will not distinguish in the following. This means that we can speak of open sets, convergent sequences, and continuous functions in normed vector spaces (i.e., with respect to the metric induced by the norm).

Furthermore, the topology induced by this canonical metric is especially compatible with the algebraic structure of the underlying vector space. Recall that two metrics are equivalent if they lead to the same convergent sequences; if the metric is induced by a norm, then $x_n \to x$ if and only if $\|x_n - x\| \to 0$.

Theorem 3.2

Let $\|\cdot\|_1$ and $\|\cdot\|_2$ be norms on the vector space X and let d_1 and d_2 be the corresponding induced metrics. Then $\|\cdot\|_1$ and $\|\cdot\|_2$ are equivalent norms if and only if d_1 and d_2 are equivalent metrics.

Proof. The equivalence of d_1 and d_2 follows directly from the equivalence of $\|\cdot\|_1$ and $\|\cdot\|_2$ together with the definition of convergence in normed vector spaces.

Assume now to the contrary that $\|\cdot\|_1$ and $\|\cdot\|_2$ are not equivalent. This implies that at least one of the inequalities in (3.1) does not hold; without loss of generality, assume that there exists no $C > 0$ such that $\|x\|_2 \leq C\|x\|_1$ for all $x \in X$. This means that we can find for every $n \in \mathbb{N}$ an $x_n \in X$ with $\|x_n\|_2 \geq n\|x_n\|_1$. Setting $y_n := (n\|x_n\|_1)^{-1}x_n$, we have $\|y_n\|_1 = \frac{1}{n} \to 0$ but $\|y_n\|_2 > 1$ for all $n \in \mathbb{N}$. Hence the sequence $\{y_n\}_{n\in\mathbb{N}}$ converges to 0 with respect to d_1 but not with respect to d_2, and therefore d_1 and d_2 cannot be equivalent, either.   □

Equivalent norms therefore lead to the same convergent sequences. Furthermore, the definition of convergence in normed vector spaces implies that they also lead to the same Cauchy sequences. In contrast to metric spaces, equivalence of norms thus conserves completeness, which accordingly is a stronger property and deserves a special name.

Definition 3.3

A complete normed vector space is called a *Banach space*.

Corollary 3.4

If $\|\cdot\|_1$ and $\|\cdot\|_2$ are equivalent norms on X, then $(X, \|\cdot\|_1)$ is a Banach space if and only if $(X, \|\cdot\|_2)$ is a Banach space.

The following useful lemma gives us another way of showing completeness of a normed vector space. Recall that a *subspace* is a subset that is closed with respect to addition of vectors and multiplication by scalars.

Lemma 3.5

Let $(X, \|\cdot\|_X)$ be a Banach space and $U \subset X$ a subspace. Then $(U, \|\cdot\|_X)$ is a Banach space if and only if U is closed.

Proof. First, it is straightforward to verify that $(U, \|\cdot\|_X)$ is a normed vector space. Let now U be closed and let $\{x_n\}_{n\in\mathbb{N}} \subset U$ be a Cauchy sequence. Since X is complete, it follows that $x_n \to x \in X$, and the closedness of U implies that $x \in U$.

Conversely, let U be complete and $\{x_n\}_{n\in\mathbb{N}} \subset U$ be a sequence with $x_n \to x \in X$. Then $\{x_n\}_{n\in\mathbb{N}}$ is in particular a Cauchy sequence (in X and hence also in U), which, due to the completeness of U, converges to some $\tilde{x} \in U$. Since limits are unique, we have $x = \tilde{x} \in U$, and hence U is closed. $\qquad\square$

Furthermore, addition of vectors, multiplication by scalars, and the norm are continuous operations.

Theorem 3.6

Let X be a normed vector space and let $\{x_n\}_{n\in\mathbb{N}}$, $\{y_n\}_{n\in\mathbb{N}} \subset X$, and $\{\lambda_n\}_{n\in\mathbb{N}} \subset \mathbb{F}$ be convergent sequences with $x_n \to x$, $y_n \to y$, and $\lambda_n \to \lambda$. Then the following properties hold:

(i) $x_n + y_n \to x + y$;

(ii) $\lambda_n x_n \to \lambda x$;

(iii) $\|x_n\| \to \|x\|$.

Proof. Properties (i) and (ii) follow as in $\mathbb{R}^n$ directly from the triangle inequality. For (iii), we use the reverse triangle inequality in the form

$$|\|x_n\| - \|x\|| = |\|x_n - x + x\| - \|x\|| \le \|x_n - x\| \to 0. \qquad\square$$

In normed vector spaces, we further have that $B_r(x) = \{y \in X : \|x - y\| \le r\}$ and hence that

$$B_r(x) = x + B_r(0) := \{y \in X : y = x + z \text{ with } z \in B_r(0)\}$$

as well as

$$B_r(0) = r B_1(0) := \{y \in X : y = rz \text{ with } z \in B_1(0)\},$$

and similarly for $U_r(x)$. It therefore usually suffices to consider the *unit ball* $B_X :=$ $B_1(0)$.

As for sequences, we define convergent series in normed vector spaces via the norm. Let $\{x_n\}_{n \in \mathbb{N}} \subset X$. A series $\sum_{n=1}^{\infty} x_n$ in X *converges* if the sequence of its partial sums $S_N := \sum_{n=1}^{N} x_n$ converges, i.e., if there exists an $x \in X$ with

$$\lim_{N \to \infty} \left\| x - \sum_{n=1}^{N} x_n \right\| = 0.$$

A series $\sum_{n=1}^{\infty} x_n$ in X *converges absolutely* if

$$\sum_{n=1}^{\infty} \|x_n\| < \infty.$$

Exactly as in $\mathbb{R}^n$, we can show the following result by looking at the sequence of partial sums of norms.

Lemma 3.7

Let X be a Banach space. If the series $\sum_{n=1}^{\infty} x_n$ is absolutely convergent, then the series is also convergent and satisfies

$$\left\| \sum_{n=1}^{\infty} x_n \right\| \le \sum_{n=1}^{\infty} \|x_n\|.$$

We now consider some canonical examples for normed vector spaces.

3.1 Finite-Dimensional Spaces

Recall that a subset V of a vector space X is called a *basis* if every $x \in X$ can be *uniquely* expressed as a linear combination $x = \sum_{v \in V} \alpha_v v$ with *coefficients* $\alpha_v \in \mathbb{F}$ for all $v \in V$. If $V = \{v_1, \ldots, v_n\}$ is finite, the number n is called the *dimension* of X. If no finite basis exists, X is infinite-dimensional.

We know from real analysis that $(\mathbb{F}, |\cdot|)$ is complete and hence a Banach space. Similarly, $\mathbb{F}^n$ is a Banach space endowed with one of the following norms:

$$\|x\|_1 := \sum_{i=1}^{n} |x_i|, \qquad \|x\|_2 := \left(\sum_{i=1}^{n} |x_i|^2 \right)^{\frac{1}{2}}, \qquad \|x\|_\infty := \max_{i=1,\ldots,n} |x_i|;$$

this follows directly from the completeness of $(\mathbb{F}, |\cdot|)$ together with the finiteness of the sums and of the set over which the maximum is taken, respectively. In each case, sequences converge if and only if they converge componentwise; these norms are therefore equivalent. This is in fact true for *all* norms on finite-dimensional spaces.

> **Theorem 3.8**
>
> *If X is a finite-dimensional vector space, then all norms on X are equivalent.*

Proof. If X is finite-dimensional, then by definition there exists a basis $\{v_1, \ldots, v_n\}$. We will show that every norm $\|\cdot\|$ is equivalent to the *Euclidean norm*

$$\|x\|_2 = \left\| \sum_{i=1}^{n} \alpha_i v_i \right\|_2 := \left(\sum_{i=1}^{n} |\alpha_i|^2 \right)^{\frac{1}{2}}.$$

A sequence converges in $(X, \|\cdot\|_2)$ if and only if the corresponding component sequences converge. In particular, $(X, \|\cdot\|_2)$ is complete since $(\mathbb{F}^n, \|\cdot\|_2)$ is complete.

Let now $M := \max\{\|v_1\|, \ldots, \|v_n\|\} > 0$. Then the triangle and Cauchy–Schwarz inequalities yield

$$\|x\| = \left\| \sum_{i=1}^{n} \alpha_i v_i \right\| \le \sum_{i=1}^{n} |\alpha_i| \|v_i\| \le \left(\sum_{i=1}^{n} |\alpha_i|^2 \right)^{\frac{1}{2}} \left(\sum_{i=1}^{n} \|v_i\|^2 \right)^{\frac{1}{2}} \le M \sqrt{n} \|x\|_2$$

and hence the second inequality in (3.1) with $C := M \sqrt{n} > 0$.

For the first inequality, we consider $S := \{x \in X : \|x\|_2 = 1\}$. Clearly, S is bounded with respect to $\|\cdot\|_2$. In addition, S is closed since S is the preimage of the closed set $\{1\}$ under the continuous function $\|\cdot\|_2$ (see Theorem 3.6 (iii)). Since we have endowed X via $\|\cdot\|_2$ with the Euclidean metric, we can apply the Heine–Borel theorem (Theorem 2.5) to conclude that S is compact. Furthermore, the already proved second inequality in (3.1) implies that the function $x \mapsto \|x\|$ is continuous with respect to $\| \cdot \|_2$ and therefore attains its minimum over S by the Weierstraß theorem (Theorem 2.9). Hence there exists an $\bar{x} \in S$ with

$$c := \|\bar{x}\| \leq \|x\| \qquad \text{for all } x \in S.$$

Since $\| \cdot \|_2$ is a norm and $0 \notin S$, it follows that $\bar{x} \neq 0$. Let now $x \in X \setminus \{0\}$ be arbitrary. Then $\frac{x}{\|x\|_2} \in S$ and hence

$$c \leq \left\| \frac{x}{\|x\|_2} \right\| = \|x\|_2^{-1} \|x\|,$$

which finally yields the first inequality. $\square$

Since completeness is conserved by passing to an equivalent norm and $(X, \| \cdot \|_2)$ is complete, we immediately obtain the following result.

> **Corollary 3.9**
> *All finite-dimensional normed vector spaces are complete.*

Finite-dimensional spaces are also special with respect to compactness. To show this, we require the following lemma.

> **Lemma 3.10 (Riesz)**
> *Let U be a closed subspace of the normed vector space X with $U \neq X$ and let $\delta \in (0, 1)$. Then there exists an $x_\delta \in X$ with $\|x_\delta\| = 1$ and*
>
> $$\|x_\delta - u\| \geq \delta \qquad \text{for all } u \in U.$$

Proof. Let $x \in X \setminus U$ be arbitrary. Since U is closed,

$$d := \inf \{\|x - u\| : u \in U\} > 0,$$

since otherwise we could find a sequence $\{u_n\}_{n \in \mathbb{N}} \subset U$ with $u_n \to x$ and $x \in \operatorname{cl} U = U$. Since $d < d/\delta$, by definition of the infimum there exists a $u_\delta \in U$ with

$$d \leq \|x - u_\delta\| < d/\delta.$$

Setting $x_\delta := \frac{x - u_\delta}{\|x - u_\delta\|}$, we have that $\|x_\delta\| = 1$.

Let now $u \in U$ be arbitrary. Since U is a subspace, $u_\delta + (\|x - u_\delta\|)u \in U$ as well, which implies that

$$\|x_\delta - u\| = \|x - u_\delta\|^{-1} \|x - u_\delta - (\|x - u_\delta\|)u\| \geq \|x - u_\delta\|^{-1} d > \delta. \qquad \square$$

We further define the *linear hull* or *span* of a (not necessarily finite) subset A of X as

$$\operatorname{lin} A = \left\{ \sum_{k=1}^{N} \lambda_k a_k : N \in \mathbb{N}, \lambda_k \in \mathbb{F}, a_k \in A \right\},$$

i.e., as the set of all *finite* linear combinations of elements from A.

Theorem 3.11

The unit ball B_X of a normed vector space X is compact if and only if X is finite-dimensional.

Proof. If X is finite-dimensional, we can show the compactness of X as in the proof of Theorem 3.8, since all topological properties such as closedness, boundedness, and compactness are conserved under equivalent norms by Theorem 3.2.

Conversely, let B_X be compact. Then there exist finitely many open balls with radius $\frac{1}{2}$ such that $B_X \subset \bigcup_{i=1}^{n} U_{\frac{1}{2}}(x_i)$ with $x_i \in B_X$. We now show that $\{x_1, \ldots, x_n\}$ is a basis of X, which is therefore finite-dimensional. Assume to the contrary that X is infinite-dimensional. Then $\operatorname{lin}\{x_1, \ldots, x_n\}$ is a proper and (since $\{x_1, \ldots, x_n\}$ is finite) closed subspace of X. By Riesz's lemma (Lemma 3.10), we can thus find an $x_{\frac{1}{2}} \in B_X$ with $\|x_{\frac{1}{2}} - x_i\| > \frac{1}{2}$ for all $i = 1, \ldots, n$, in contradiction to the choice of the x_i. $\qquad \square$

This implies that closed bounded sets in infinite-dimensional normed vector spaces are *not* necessarily compact; the lack of this useful property was a significant influence on the development of novel functional-analytic methods.

3.2 Sequence Spaces

We now consider the simplest examples of infinite-dimensional normed vector spaces. Let

$$\mathbb{F}^{\mathbb{N}} := \left\{ \{x_k\}_{k \in \mathbb{N}} : x_k \in \mathbb{F} \text{ for all } k \in \mathbb{N} \right\}$$

denote the set of all sequences in $\mathbb{F}$ and define the subsets

$$\ell^\infty(\mathbb{F}) := \left\{ x \in \mathbb{F}^\mathbb{N} : x = \{x_k\}_{k\in\mathbb{N}} \text{ is bounded} \right\},$$

$$c(\mathbb{F}) := \left\{ x \in \mathbb{F}^\mathbb{N} : x = \{x_k\}_{k\in\mathbb{N}} \text{ is convergent} \right\},$$

$$c_0(\mathbb{F}) := \left\{ x \in \mathbb{F}^\mathbb{N} : x = \{x_k\}_{k\in\mathbb{N}} \text{ is a null sequence} \right\},$$

$$c_e(\mathbb{F}) := \left\{ x \in \mathbb{F}^\mathbb{N} : x = \{x_k\}_{k\in\mathbb{N}} \text{ is a finite sequence} \right\},$$

i.e., $\{x_k\}_{k\in\mathbb{N}} \in c_e(\mathbb{F})$ if and only if there exists an $N \in \mathbb{N}$ with $x_k = 0$ for all $k \geq N$. It is straightforward to verify that these sets are closed with respect to componentwise addition of vectors and multiplication by scalars and that they form vector spaces. We now endow these spaces with the *supremum norm*

$$\|x\|_\infty := \sup_{k\in\mathbb{N}} |x_k| \qquad \text{for } x = \{x_k\}_{k\in\mathbb{N}}.$$

Theorem 3.12

The space $(\ell^\infty(\mathbb{F}), \|\cdot\|_\infty)$ is a Banach space.

Proof. We first verify that this indeed defines a normed vector space. Note that by definition, $x \in \mathbb{F}^\mathbb{N}$ is bounded if and only if $\|x\|_\infty$ is finite. Nondegeneracy and homogeneity of the norm are also clear. Let now $x, y \in \ell^\infty(\mathbb{F})$ and $n \in \mathbb{N}$ be arbitrary. Then

$$|x_k + y_k| \leq |x_k| + |y_k| \leq \sup_{k\in\mathbb{N}} |x_k| + \sup_{k\in\mathbb{N}} |y_k| = \|x\|_\infty + \|y\|_\infty,$$

and taking the supremum over all $k \in \mathbb{N}$ yields the triangle inequality.

To show completeness, we have to consider sequences of sequences; for this we again change notation and write $x = \{x(k)\}_{k\in\mathbb{N}} \in \mathbb{F}^\mathbb{N}$. Let now $\{x_n\}_{n\in\mathbb{N}}$ be a Cauchy sequence in $(\ell^\infty(\mathbb{F}), \|\cdot\|_\infty)$. Since $|x_n(k)| \leq \|x_n\|_\infty$ for all $k \in \mathbb{N}$, this implies that $\{x_n(k)\}_{n\in\mathbb{N}}$ is a Cauchy sequence, which by the completeness of $\mathbb{F}$ has a limit $x(k) \in \mathbb{F}$. This defines a sequence $x := \{x(k)\}_{k\in\mathbb{N}}$, for which we have to show that $x \in \ell^\infty(\mathbb{F})$ and $\|x_n - x\|_\infty \to 0$. Since $\{x_n\}_{n\in\mathbb{N}}$ is a Cauchy sequence, there exists for every $\varepsilon > 0$ an $N \in \mathbb{N}$ with

$$|x_n(k) - x_m(k)| \leq \|x_n - x_m\|_\infty \leq \varepsilon \qquad \text{for all } n, m \geq N, \ k \in \mathbb{N}.$$

Let now $k \in \mathbb{N}$ be arbitrary. Since $x_n(k) \to x(k)$, there further exists an $M(k)$ with

$$|x_M(k) - x(k)| \leq \varepsilon,$$

where we can assume without loss of generality that $M(k) \geq N$. Hence for all $n \geq N$,

$$|x_n(k) - x(k)| \leq |x_n(k) - x_{M(k)}(k)| + |x_{M(k)}(k) - x(k)| \leq 2\varepsilon.$$

On the one hand, this implies that

$$|x(k)| \leq |x_N(k)| + |x(k) - x_N(k)| \leq \|x_N\|_\infty + 2\varepsilon < \infty$$

and hence that $x \in \ell^\infty(\mathbb{F})$. On the other hand, by taking the supremum over all $k \in \mathbb{N}$, we obtain that

$$\|x_n - x\|_\infty \leq 2\varepsilon \qquad \text{for all } n \geq N$$

and hence that $x_n \to x$ with respect to the supremum norm. $\qquad\square$

For the remaining spaces, we use Lemma 3.5.

> **Theorem 3.13**
>
> *Endowed with the supremum norm, $c(\mathbb{F})$ and $c_0(\mathbb{F})$ are Banach spaces. The space $c_e(\mathbb{F})$ is not compact.*

Proof. It is not hard to see that all three spaces are subspaces of $\ell^\infty(\mathbb{F})$ and therefore normed vector spaces when endowed with the supremum norm. Hence it remains only to show that $c(\mathbb{F})$ and $c_0(\mathbb{F})$ are closed but not $c_e(\mathbb{F})$.

Let first $\{x_n\}_{n\in\mathbb{N}} \subset c(\mathbb{F})$ be a sequence converging in $\ell^\infty(\mathbb{F})$ to a limit $x \in \ell^\infty(\mathbb{F})$. We now show that $x = \{x(k)\}_{k\in\mathbb{N}}$ is itself a Cauchy sequence (in $\mathbb{F}$). Let $\varepsilon > 0$ be given. By the convergence of the sequence $\{x_n\}_{n\in\mathbb{N}}$, there then exists an $N \in \mathbb{N}$ with $\|x_N - x\|_\infty \leq \varepsilon$. Since the sequence $x_N = \{x_N(k)\}_{k\in\mathbb{N}} \in c(\mathbb{F})$ is convergent and hence a Cauchy sequence, there further exists an $M \in \mathbb{N}$ with $|x_N(k) - x_N(l)| \leq \varepsilon$ for all $k, l \geq M$. We thus obtain for all $k, l \geq M$ that

$$|x(k) - x(l)| \leq |x(k) - x_N(k)| + |x_N(k) - x_N(l)| + |x_N(l) - x(l)|$$

$$\leq 2\|x - x_N\|_\infty + \varepsilon \leq 3\varepsilon,$$

i.e., x is a Cauchy sequence and hence $x \in c(\mathbb{F})$.

Let now $\{x_n\}_{n\in\mathbb{N}} \subset c_0(\mathbb{F})$ be a sequence converging in $\ell^\infty(\mathbb{F})$ to a limit $x \in \ell^\infty(\mathbb{F})$. We already know that $x \in c(\mathbb{F})$ and have only to prove that $\lim_{k\to\infty} x(k) = 0$. By proceeding as before, we obtain for every $\varepsilon > 0$ an $N \in \mathbb{N}$ and an $M \in \mathbb{N}$ such that

$$|x(M)| \leq |x(M) - x_N(M)| + |x_N(M)| \leq \|x - x_N\|_\infty + |x_N(M)| \leq \varepsilon + \varepsilon$$

since $x_N \in c_0(\mathbb{F})$. Hence x is a null sequence as well.

For $c_e(\mathbb{F})$, consider for $n, k \in \mathbb{N}$

$$x_n(k) := \begin{cases} \frac{1}{k} & \text{if } k \le n, \\ 0 & \text{otherwise,} \end{cases} \qquad x(k) := \frac{1}{k}.$$

Then $x_n \in c_e(\mathbb{F})$ for all $n \in \mathbb{N}$ and $\|x_n - x\|_\infty = \frac{1}{n+1} \to 0$, but $x \notin c_e(\mathbb{F})$. $\square$

Proceeding as in the last step, one can show that $c_e(\mathbb{F})$ is dense in $c_0(\mathbb{F})$ (consider for given $x \in c_0(\mathbb{F})$ the sequence $\{x_n\}_{n \in \mathbb{N}} \subset c_e(\mathbb{F})$ with $x_n(k) = x(k)$ for $k \le n$ and $x_n(k) = 0$ otherwise).

Another class of Banach spaces is obtained using the *p-norms*. For $x = \{x_k\}_{k \in \mathbb{N}} \in \mathbb{F}^{\mathbb{N}}$, we define

$$\|x\|_p := \left(\sum_{k \in \mathbb{N}} |x_k|^p \right)^{\frac{1}{p}}, \qquad 1 \le p < \infty,$$

as well as

$$\ell^p(\mathbb{F}) := \left\{ x \in \mathbb{F}^{\mathbb{N}} : \|x\|_p < \infty \right\}.$$

Theorem 3.14
For all $1 \le p < \infty$, $\ell^p(\mathbb{F})$ is a Banach space.

Proof. Again, it is straightforward to verify the nondegeneracy and the homogeneity of the mapping $\| \cdot \|_p$. For the triangle inequality, we use the *Minkowski inequality*

$$\left(\sum_{k=1}^{N} |x_k + y_k|^p \right)^{\frac{1}{p}} \le \left(\sum_{k=1}^{N} |x_k|^p \right)^{\frac{1}{p}} + \left(\sum_{k=1}^{N} |y_k|^p \right)^{\frac{1}{p}}$$

to obtain

$$\sum_{k=1}^{N} |x_k + y_k|^p \le (\|x\|_p + \|y\|_p)^p \qquad \text{for all } N \in \mathbb{N}.$$

Passing to the limit $N \to \infty$ and taking the pth root then yields that

$$\|x + y\|_p \le \|x\|_p + \|y\|_p.$$

Hence $x + y \in \ell^p(\mathbb{F})$, and therefore $\ell^p(\mathbb{F})$ is a normed vector space.

Let now $\{x_n\}_{n\in\mathbb{N}} \subset \ell^p(\mathbb{F})$ be a Cauchy sequence. Then,

$$|x_n(k) - x_m(k)|^p \leq \sum_{j=1}^{\infty} |x_n(j) - x_m(j)|^p = \|x_n - x_m\|_p^p \qquad \text{for all } k, n, m \in \mathbb{N}.$$

Hence $\{x_n(k)\}_{n\in\mathbb{N}}$ is a Cauchy sequence for all $k \in \mathbb{N}$ and therefore converges to some $x(k) \in \mathbb{F}$. We can thus find for all $\varepsilon > 0$ an $M \in \mathbb{N}$ such that

$$\|x_n - x_m\|_p \leq \varepsilon \qquad \text{for all } m, n \geq M,$$

as well as for all $N \in \mathbb{N}$ an $m = m(N) \geq N$ such that

$$\left(\sum_{k=1}^{N} |x_m(k) - x(k)|^p\right)^{\frac{1}{p}} \leq \varepsilon$$

(this is possible since $x_n(k) \to x(k)$ for all $k \leq N$). For all $n \geq M$, we then have (again using the Minkowski inequality) that

$$\left(\sum_{k=1}^{N} |x_n(k) - x(k)|^p\right)^{\frac{1}{p}} \leq \left(\sum_{k=1}^{N} |x_n(k) - x_m(k)|^p\right)^{\frac{1}{p}} + \left(\sum_{k=1}^{N} |x_m(k) - x(k)|^p\right)^{\frac{1}{p}}$$

$$\leq \|x_n - x_m\|_p + \left(\sum_{k=1}^{N} |x_m(k) - x(k)|^p\right)^{\frac{1}{p}}$$

$$\leq \varepsilon + \varepsilon.$$

Passing to the limit $N \to \infty$ yields that $\|x_n - x\|_p \leq 2\varepsilon$ for all $n \geq M$ and hence that $x_n \to x$ and $x_n - x \in \ell^p(\mathbb{F})$, which implies that $x = (x - x_n) + x_n \in \ell^p(\mathbb{F})$. $\square$

As in the proof of Theorem 3.13, one can show that $c_e(\mathbb{F})$ is dense in $\ell^p(\mathbb{F})$ for $1 \leq p < \infty$.

Theorem 3.15
The spaces $c_0(\mathbb{F})$ and $\ell^p(\mathbb{F})$ for $1 \leq p < \infty$ are separable. The space $\ell^\infty(\mathbb{F})$ is not separable.

Proof. For $c_0(\mathbb{F})$ and $\ell^p(\mathbb{F})$, consider for $\mathbb{F} = \mathbb{R}$ the space $c_e(\mathbb{Q})$ of rational finite sequences and for $\mathbb{F} = \mathbb{C}$ the space $c_e(\mathbb{Q} + i\mathbb{Q})$. Since the rational numbers are dense in $\mathbb{R}$, a diagonal sequence argument shows that these sets are dense in $c_0(\mathbb{F})$ and $\ell^p(\mathbb{F})$. Furthermore, these sets are countable, which implies the claimed separability.

For $\ell^\infty(\mathbb{F})$, consider an arbitrary subset $M \subset \mathbb{N}$ and define $x_M \in \ell^\infty(\mathbb{F})$ via

$$x_M(k) := \begin{cases} 1 & \text{if } k \in M, \\ 0 & \text{otherwise.} \end{cases}$$

For all $M, N \subset \mathbb{N}$ with $M \neq N$, we then have that $\|x_M - x_N\|_\infty = 1$. Let now $A \subset \ell^\infty(\mathbb{F})$ be an arbitrary countable subset. Then for all $x \in A$, the open ball $U_{\frac{1}{2}}(x)$ can contain at most one such x_M. But since the set of all subsets of $\mathbb{N}$ and thus the number of such x_M is uncountable, A cannot be dense.   □

3.3   Function Spaces

The space of $\mathbb{F}$-valued functions also defines a vector space over $\mathbb{F}$ if addition of functions and multiplication by scalars is defined pointwise. If (X, d) is a metric space, we define the space of bounded functions on X,

$$B(X) := \{f : X \to \mathbb{F} : f \text{ bounded}\},$$

as well as the *supremum norm*

$$\|f\|_\infty := \sup_{x \in X} |f(x)|.$$

Exactly as in the proof of Theorem 3.12 (simply replacing $k \in \mathbb{N}$ with $x \in X$), one shows that $(B(X), \| \cdot \|_\infty)$ is a Banach space. Similarly, one proves that $C_b(X)$ is a closed subspace of $B(X)$ and hence that $(C_b(X), \| \cdot \|_\infty)$ is a Banach space as well; this also follows from the fact that $\| \cdot \|_\infty$ induces the metric in (1.2) and hence the topology of uniform convergence together with the fact that the limit of *uniformly* convergent sequences of continuous functions is again continuous. The function space corresponding to $c_0(\mathbb{F})$ is the space of functions vanishing at infinity,

$$C_0(X) := \{f \in C(X) : \text{for all } \varepsilon > 0, \text{ the set } \{x \in X : |f(x)| \geq \varepsilon\} \text{ is compact}\},$$

which in turn is a closed subspace of $C_b(X)$ and hence a Banach space together with the supremum norm.

By the Weierstraß approximation theorem,[1] every continuous function can be approximated arbitrarily well by polynomials. A diagonal sequence argument as in the proof of Theorem 3.15 then shows that for $X \subset \mathbb{R}^n$, polynomials with rational coefficients are dense in $C_b(X)$ and hence that $C_b(X)$ is separable.

---

[1] See, e.g., [20, Theorem 7.26].

It is also possible to define function spaces corresponding to $\ell^p(\mathbb{F})$, $1 \le p \le \infty$. The construction is involved and requires significant measure-theoretic groundwork; we therefore treat only these spaces in a cursory manner (in particular since the genuinely functional-analytic arguments are essentially the same as for $\ell^p(\mathbb{F})$) and refer for details to [3, Chapter 4] or [5, Chapters V and VI].

Let $\Omega \subset \mathbb{R}^n$ be a Lebesgue-measurable[2] subset and define for a Lebesgue-measurable function $f : \Omega \to \mathbb{R}$

$$\|f\|_p := \left(\int_\Omega |f(x)|^p \, dx \right)^{\frac{1}{p}}, \qquad 1 \le p < \infty,$$

$$\|f\|_\infty := \operatorname*{ess\,sup}_{x \in \Omega} |f(x)|,$$

where the *essential supremum* of $|f|$ is defined as

$$\operatorname*{ess\,sup}_{x \in \Omega} |f(x)| := \inf \{M > 0 : \{x \in \Omega : |f(x)| > M\} \text{ has Lebesgue measure } 0\}.$$

In the following, we combine functions that only differ from f on a set of Lebesgue measure 0 into a single equivalence class, which is again denoted by f. Then for all $1 \le p \le \infty$, the space

$$L^p(\Omega) := \left\{ f : \Omega \to \mathbb{R} : \|f\|_p < \infty \right\}$$

together with the corresponding norm is a Banach space.[3] It is possible to show that $C_b(\Omega)$ is dense in $L^p(\Omega)$ for $1 \le p < \infty$, which implies that these spaces are separable. In contrast, $L^\infty(\Omega)$ is not separable, which can be shown by similar arguments as for $\ell^\infty(\Omega)$.

Problems

Problem 3.1 *(Convexity of the unit ball)*
Let X be a vector space over $\mathbb{R}$ or $\mathbb{C}$ and let $p : X \to [0, \infty)$ be a mapping with the following properties:

(i) $p(x) = 0$ if and only if $x = 0$ (nondegeneracy);
(ii) $p(\lambda x) = |\lambda| p(x)$ for all $\lambda \in \mathbb{F}$ and all $x \in X$ (homogeneity).

---

[2]See, e.g., [5, Chapters II and III] for this and the following concepts from measure theory.

[3]This construction also works in more general measure spaces, which is important for probability theory; see, e.g., [3, Theorem 4.8] or [5, Proposition 6.3].

Show that p is a norm if and only if the *p-unit ball*

$$\{x \in X : p(x) \le 1\}$$

is convex.

Recall that a set $U \subset X$ is convex if $\lambda x + (1 - \lambda)y \in U$ for all $x, y \in U$ and all $\lambda \in [0, 1]$.

Problem 3.2 *(Normed vector spaces)*
For $x = \{x_n\}_{n \in \mathbb{N}} \in \ell^1(\mathbb{R})$, set

$$\|x\| = \sup_{n \in \mathbb{N}} \left| \sum_{k=1}^{n} x_k \right|.$$

Show that or give a counterexample for:

(i) $\left(\ell^1(\mathbb{R}), \|\cdot\|\right)$ is a normed vector space;
(ii) $\left(\ell^1(\mathbb{R}), \|\cdot\|\right)$ is a Banach space;
(iii) $\|\cdot\|$ is equivalent to $\|\cdot\|_1$.

Problem 3.3 *(Separable normed vector spaces)*
Let X be a normed vector space over $\mathbb{R}$. Show that X is separable if and only if there exists a separable set $A \subset X$ with $X = \mathrm{cl}(\mathrm{lin}\, A)$.

Problem 3.4 *(Subspaces of ℓ^p)*
Let $1 \le p < \infty$ and

$$G_p = \left\{ \{x_k\}_{k \in \mathbb{N}} \in \ell^p(\mathbb{R}) : \sum_{k=1}^{\infty} x_k = 0 \right\}.$$

Show that

(i) the set G_p is a subspace of $\ell^p(\mathbb{R})$;
(ii) for $1 < p < \infty$, the set G_p is not closed;
(iii) for $p = 1$, the set G_p is closed.
   Hint: G_p is the preimage of $\{0\}$ under a certain mapping.

Problem 3.5 *(Compact subsets of ℓ^p)*
Let $1 \le p < \infty$ and $A \subset \ell^p(\mathbb{F})$. Show that the following properties are equivalent:

(i) A is relatively compact;
(ii) A is bounded and

$$\lim_{n \to \infty} \sup_{x \in A} \left(\sum_{k=n}^{\infty} |x(k)|^p \right)^{\frac{1}{p}} = 0.$$

Hint: Follow the proof of the Arzelà–Ascoli theorem (Theorem 2.11) and argue by contradiction.

Problem 3.6 *(Series in normed vector spaces)*

Show that X is complete if and only if every absolutely convergent series in X converges.

Hint: Use that a Cauchy sequence that contains a convergent subsequence must itself converge.

Linear Operators

4

We now study mappings between normed vector spaces, exploiting again the interplay between algebraic and topological properties. For normed vector spaces $(X, \|\cdot\|_X)$ and $(Y, \|\cdot\|_Y)$, we thus consider mappings $T : X \to Y$ that are

(i) *linear*, i.e., $T(\lambda x_1 + x_2) = \lambda T(x_1) + T(x_2)$ for every $x_1, x_2 \in X, \lambda \in \mathbb{F}$;
(ii) continuous in the sense of Definition 1.7, i.e., for which $x_n \to x$ implies that $T x_n \to T x$.

A mapping with these properties is also called a *continuous linear operator*; to stress the linearity, one often writes $Tx := T(x)$. As in linear algebra, we define for a linear operator $T : X \to Y$

(i) the *kernel* or *null space* $\ker T := T^{-1}(\{0\}) = \{x \in X : Tx = 0\} \subset X$;
(ii) the *range* $\operatorname{ran} T := T(X) = \{Tx : x \in X\} \subset Y$;
(iii) the *graph* $\operatorname{graph} T := \{(x, Tx) : x \in X\} \subset X \times Y$.

The linearity of T directly implies that these are subspaces. If T is continuous, $\ker T$ is closed as the preimage of the closed set $\{0\}$. However, this is not necessarily the case for $\operatorname{ran} T$ (or $\operatorname{graph} T$), which is a fundamental difficulty in dealing with mappings between infinite-dimensional vector spaces.

The additional linearity allows a simpler characterization of continuity.

© Springer Nature Switzerland AG 2020

C. Clason, *Introduction to Functional Analysis*, Compact Textbooks in Mathematics, https://doi.org/10.1007/978-3-030-52784-6_4

Lemma 4.1
*Let $T : X \to Y$ be a linear mapping between the normed vector spaces $(X, \| \cdot \|_X)$ and
$(Y, \| \cdot \|_Y)$. Then the following properties are equivalent:*

  (i) T is continuous on X;
  (ii) T is continuous at $0 \in X$;
  (iii) T is bounded, i.e., there exists a constant $C \geq 0$ such that

$$\|Tx\|_Y \leq C\|x\|_X \qquad \text{for all } x \in X.$$

Proof. (i) $\Rightarrow$ (ii) is clear.

  (ii) $\Rightarrow$ (iii): By Definition 1.7 (ii), we can find for $\varepsilon = 1$ some $\delta > 0$ such that $T(B_\delta(0)) \subset B_1(T(0)) = B_1(0)$. The definition of closed balls in normed vector spaces then implies that

$$\|Tx\|_Y \leq 1 \qquad \text{for all } x \in X \text{ with } \|x\|_X \leq \delta. \tag{4.1}$$

We now have for all $x \in X \setminus \{0\}$ that $\delta \frac{x}{\|x\|_X} \in B_\delta(0)$, and hence (4.1) can be reformulated using the linearity of T and the homogeneity of the norm as

$$\|Tx\|_Y \leq \frac{1}{\delta}\|x\|_X.$$

This yields the claim with $C := \delta^{-1}$.

  (iii) $\Rightarrow$ (i): Let $x \in X$ and $\varepsilon > 0$ be given. We now show that there exists a $\delta > 0$ such that $T(B_\delta(x)) \subset B_\varepsilon(Tx)$. Specifically, we choose $\delta := \frac{\varepsilon}{C}$; then we have for all $z \in B_\delta(x)$ that

$$\|Tz - Tx\|_Y = \|T(z - x)\|_Y \leq C\|z - x\|_X \leq C\delta = \varepsilon,$$

which yields the claim. $\square$

On finite-dimensional normed vector spaces, all linear operators are continuous.

Lemma 4.2
Let X and Y be normed vector spaces and $T : X \to Y$ a linear operator. If X is finite-dimensional, then T is continuous.

Proof. Since X is finite-dimensional, there exists a basis $\{v_1, \ldots, v_n\}$ of X. For given

$$x = \sum_{k=1}^{n} x_k v_k \in X, \qquad \text{with } x_k \in \mathbb{F} \quad \text{for all} \quad k = 1, \ldots, n,$$

we thus have

$$\|Tx\|_Y \leq \sum_{k=1}^{n} |x_k| \|T v_k\|_Y \leq \max_{k=1,\ldots,n} \|T v_k\|_Y \sum_{k=1}^{n} |x_k| =: M \|x\|_1$$

with $M := \max_{k=1,\ldots,n} \|T v_k\|_Y < \infty$. Since by Theorem 3.8, all norms on a finite-dimensional vector space are equivalent, there exists a $C > 0$ such that $\|x\|_1 \leq C \|x\|_X$ for all $x \in X$. This shows the claimed continuity. □

The set of all continuous linear operators from X to Y is denoted by $L(X, Y)$; this set becomes a vector space by pointwise addition and scalar multiplication, i.e., by defining $T_1 + \alpha T_2 \in L(X, Y)$ for $T_1, T_2 \in L(X, Y)$ and $\alpha \in \mathbb{F}$ via $(T_1 + \alpha T_2) x := T_1 x + \alpha T_2 x$ for all $x \in X$. A reasonable choice for the norm of a continuous linear operator is the smallest possible constant in Lemma 4.1 (iii). Specifically, we define the *operator norm*

$$\|T\|_{L(X,Y)} := \sup_{x \in B_X} \|Tx\|_Y. \tag{4.2}$$

The justification for this definition is provided by the following lemma, which can be proved by simple estimation or inserting suitable choices of $x \in X$.

Lemma 4.3
For all $T \in L(X, Y)$, the following relations hold:

(i) $\|T\|_{L(X,Y)} = \sup_{x \in X, \|x\|_X < 1} \|Tx\|_Y = \sup_{x \in X, \|x\|_X = 1} \|Tx\|_Y = \sup_{x \in X \setminus \{0\}} \dfrac{\|Tx\|_Y}{\|x\|_X}$;

(ii) $\|T\|_{L(X,Y)} = \inf \{ C > 0 : \|Tx\|_Y \leq C \|x\|_X \text{ for all } x \in X \}$;

(iii) $\|Tx\|_Y \leq \|T\|_{L(X,Y)} \|x\|_X \quad$ *for all $x \in X$.*

Clearly, the *identity* $\mathrm{Id}_X : X \rightarrow X, x \mapsto x$, on any normed vector space is a continuous linear operator with operator norm 1. Furthermore, the definition coincides with the induced matrix norm known from numerical linear algebra; the most important example is the *spectral norm* induced by the Euclidean norm (i.e., $X = (\mathbb{R}^n, \| \cdot \|_2)$). The following examples illustrate the definition in infinite-dimensional spaces.

Example 4.4

(i) Let $X = c(\mathbb{F})$ and $T : X \to \mathbb{F}$, $x = \{x_k\}_{k \in \mathbb{N}} \mapsto \lim_{k \to \infty} x_k$. The linearity of T is an immediate consequence of the calculus for convergent sequences. To show the continuity, consider

$$|Tx| = \left| \lim_{k \to \infty} x_k \right| = \lim_{k \to \infty} |x_k| \le \sup_{k \in \mathbb{N}} |x_k| = \|x\|_\infty.$$

Since the constant sequence $x = \{1\}_{k \in \mathbb{N}}$ satisfies $|Tx| = 1 = \|x\|_\infty$, we have $\|T\|_{L(X,\mathbb{F})} = 1$.[1]

(ii) Analogously, one shows that for $X = C([0, 1])$, the *point evaluation* $T : X \to \mathbb{R}$, $x \mapsto x(0)$, is linear and continuous with operator norm $\|T\|_{L(C, \mathbb{R})} = 1$.

(iii) Let $X = C([a, b])$ and $T : X \to \mathbb{R}$, $x \mapsto \int_a^b x(t)\, dt$. Then T is also linear and continuous, since

$$|Tx| = \left| \int_a^b x(t)\, dt \right| \le (b - a)\|x\|_\infty,$$

with equality for constant functions. Hence $\|T\|_{L(C, \mathbb{R})} = b - a$.

(iv) Let $X = C^1([0, 1]) := \{f : [0, 1] \to \mathbb{R} : f \text{ continuously differentiable}\}$ and $Y = C([0, 1])$, and consider the *derivative operator* $D : X \to Y$, $x \mapsto x'$. The linearity again follows immediately from the definition. If we endow $C([0, 1])$ and $C^1([0, 1])$ with the supremum norm, then D is *not* continuous, since $x_n(t) := t^n$ satisfies $\|x_n\|_\infty = 1$, but $\|Dx_n\|_\infty = \|n t^{n-1}\|_\infty = n$.

On the other hand, if we endow $C^1([0, 1])$ with the norm $\|x\|_{C^1} := \|x\|_\infty + \|x'\|_\infty$, then D is continuous with operator norm 1 since $\|Dx\|_\infty = \|x'\|_\infty \le \|x\|_{C^1}$.

It remains to show that (4.2) indeed defines a norm.

Theorem 4.5

The pair $(L(X, Y), \|\cdot\|_{L(X,Y)})$ is a normed vector space. If Y is complete, then $L(X, Y)$ is a Banach space.

Proof. Lemma 4.3 (ii) implies that for every continuous linear operator $T : X \to Y$, we have that $\|T\|_{L(X,Y)} < \infty$. The homogeneity and nondegeneracy follow from the corresponding properties of the norm on Y, the latter using the fact that $T = 0$ if and only if $Tx = 0$ for all $x \in X$. For the triangle inequality, let $x \in B_X$ be arbitrary. Then we obtain from

---

[1] Since $c_0(\mathbb{F}) = \ker T$, the continuity of T together with Corollary 1.9 yields a much more elegant proof of the closedness of $c_0(\mathbb{F})$ in $c(\mathbb{F})$; cf. Theorem 3.13.

Lemma 4.3 (iii) for all $S, T \in L(X, Y)$ that

$$\|(S + T)x\|_Y = \|Sx + Tx\|_Y \leq \|Sx\|_Y + \|Tx\|_Y \leq \|S\|_{L(X,Y)} + \|T\|_{L(X,Y)},$$

and taking the supremum over all $x \in B_X$ yields the claim.

For the completeness, let $\{T_n\}_{n \in \mathbb{N}}$ be a Cauchy sequence in $L(X, Y)$. Then for all $x \in X$, the sequence $\{T_n x\}_{n \in \mathbb{N}}$ is also a Cauchy sequence in Y and therefore converges by assumption to some $y_x \in Y$. This defines a mapping $T : X \to Y, x \mapsto y_x$. This mapping is linear since

$$T(\lambda x_1 + x_2) = \lim_{n \to \infty} T_n(\lambda x_1 + x_2) = \lim_{n \to \infty} \lambda T_n x_1 + \lim_{n \to \infty} T_n x_2 = \lambda T(x_1) + T(x_2),$$

where we have used the linearity of T_n as well as the continuity of addition and scalar multiplication (Theorem 3.6 (i), (ii)).

We now show that $\|T\|_{L(X,Y)} < \infty$ (and hence that $T \in L(X, Y)$) as well as that $\|T_n - T\|_{L(X,Y)} \to 0$. Since $\{T_n\}_{n \in \mathbb{N}}$ is a Cauchy sequence, we can find for every $\varepsilon > 0$ an $N \in \mathbb{N}$ such that

$$\|T_n - T_m\|_{L(X,Y)} \leq \varepsilon \qquad \text{for all } n, m \geq N.$$

Let now $x \in B_X$ be arbitrary. Since $T_n x \to Tx$, we can find an $M = M(\varepsilon, x) \geq N$ with

$$\|T_M x - Tx\|_Y \leq \varepsilon.$$

Taking these together, we obtain that

$$\|T_n x - Tx\|_Y \leq \|T_n x - T_M x\|_Y + \|T_M x - Tx\|_Y \leq \|T_n - T_m\|_{L(X,Y)} + \varepsilon \leq 2\varepsilon.$$

Taking the supremum over all $x \in B_X$ then yields $\|T_n - T\|_{L(X,Y)} \leq 2\varepsilon$. On the one hand, we obtain from this via the triangle inequality that $\|T\|_{L(X,Y)} \leq \|T_n\|_{L(X,Y)} + 2\varepsilon < \infty$. On the other hand, since $\varepsilon > 0$ was arbitrary, this implies that $\|T_n - T\|_{L(X,Y)} \to 0$. $\square$

We next show two useful properties. First, an immediate consequence of Lemma 4.3 (iii) is

Corollary 4.6

Let X, Y, Z be normed vector spaces, $T \in L(X, Y)$, and $S \in L(Y, Z)$. Then $S \circ T \in L(X, Z)$ with $\|S \circ T\|_{L(X,Z)} \leq \|S\|_{L(Y,Z)} \|T\|_{L(X,Y)}$.

The following result is helpful for the construction of continuous linear operators with given properties.

Theorem 4.7

Let $U \subset X$ be a dense subspace, Y a Banach space, and $T \in L(U, Y)$. Then there exists a unique continuous extension $S \in L(X, Y)$ with $S|_U = T$ and $\|S\|_{L(X,Y)} = \|T\|_{L(U,Y)}$.

Proof. Let $x \in X$. By assumption, there exists a sequence $\{x_n\}_{n \in \mathbb{N}} \subset U$ with $x_n \to x$. In particular, $\{x_n\}_{n \in \mathbb{N}}$ is a Cauchy sequence. Since

$$\|T x_n - T x_m\|_Y \leq \|T\|_{L(U,Y)} \|x_n - x_m\|_X \qquad n, m \in \mathbb{N}, \tag{4.3}$$

$\{T x_n\}_{n \in \mathbb{N}}$ is also a Cauchy sequence that converges in the Banach space Y to some $y_x \in Y$. As before, the mapping $S : X \to Y$, $x \mapsto y_x$, defines a linear operator. The fact that this mapping is unique (i.e., does not depend on the specific choice of the Cauchy sequence) can be seen as follows: If two Cauchy sequences $\{x_n\}_{n \in \mathbb{N}}$ and $\{\tilde{x}_n\}_{n \in \mathbb{N}}$ converge to the same limit x, then $\{x_n - \tilde{x}_n\}_{n \in \mathbb{N}}$ is a null sequence, and (4.3) implies that $T x_n - T \tilde{x}_n \to 0$ as well.

With the same choice of $x_n \to x$, we can now use the continuity of S and of the norm (Theorem 3.6 (iii)) together with Lemma 4.3 (iii) to obtain that

$$\|Sx\|_Y = \lim_{n \to \infty} \|Sx_n\|_Y = \lim_{n \to \infty} \|T x_n\|_Y$$

$$\leq \|T\|_{L(U,Y)} \left(\lim_{n \to \infty} \|x_n\|_X \right) = \|T\|_{L(U,Y)} \|x\|_X \qquad \text{for all } x \in X.$$

Taking the supremum over all $x \in B_X$ then yields $\|S\|_{L(X,Y)} \leq \|T\|_{L(U,Y)}$. The reverse inequality follows from

$$\|T\|_{L(U,Y)} = \sup_{x \in B_U} \|T x\|_Y = \sup_{x \in B_U} \|Sx\|_Y \leq \sup_{x \in B_X} \|Sx\|_Y = \|S\|_{L(X,Y)},$$

since $B_U \subset B_X$ and hence the supremum is taken over a larger set. $\qquad \square$

The proof is a classical example of a *density argument* that is typical in functional analysis: to prove that a property holds for all elements of X, it suffices to show that it holds on a dense subset and that it is preserved on passing to the limit.

Continuous linear operators between normed vector spaces allow transferring properties from one to the other.

We recall from linear algebra that an operator $T : X \to Y$ is called *injective* if $\ker T = \{0\}$ and *surjective* if $\operatorname{ran} T = Y$. An operator that is both injective and surjective is called *bijective*; in this case we can define a linear operator $T^{-1} : Y \to X$ via the one-to-one mapping $y \mapsto x \in T^{-1}(\{y\})$, which satisfies $T^{-1}T = \operatorname{Id}_X$ as well as $T T^{-1} = \operatorname{Id}_Y$. We then call T^{-1} the *inverse* and T *invertible*. If $T^{-1} \in L(Y, X)$ (which need not be the case!), then T is called *continuously invertible*.

A continuously invertible operator is also called an *isomorphism*. If $\|Tx\|_Y = \|x\|_X$ for all $x \in X$, then T is called an *isometry*.

We now call two vector spaces X and Y *isomorphic* and write $X \simeq Y$ if there exists an isomorphism $T : X \to Y$. Analogously, we call X and Y *isometrically isomorphic* and write $X \cong Y$ if there exists a $T \in L(X, Y)$ that is both an isomorphism and an isometry. Isometric vector spaces are in essence only different representations of the same space; for isometrically isomorphic spaces, these representations hold "uniformly". Similarly to Corollary 3.4, if X is a Banach space and $X \simeq Y$, then Y is a Banach space as well.

If two norms $\| \cdot \|_1$ and $\| \cdot \|_2$ are equivalent on X, then the identity $\mathrm{Id}_X : (X, \| \cdot \|_1) \to (X, \| \cdot \|_2)$ is an isomorphism. Different isomorphic spaces are thus a generalization of equivalent norms on the same space. Generalizing Theorem 3.8 accordingly, finite-dimensional spaces of the same dimension are always (but not necessarily isometrically) isomorphic.

Theorem 4.8

If X and Y are finite-dimensional normed vector spaces with $\dim(X) = \dim(Y)$, then $X \simeq Y$.

Proof. We show that every n-dimensional normed vector space $(X, \| \cdot \|)$ is isomorphic to $(\mathbb{F}^n, \| \cdot \|_2)$. Let $\{v_1, \ldots, v_n\}$ be a basis of X. Then the operator

$$T : X \to \mathbb{F}^n, \qquad x = \sum_{k=1}^n x_k v_k \mapsto (x_1, \ldots, x_n),$$

as well as the inverse

$$T^{-1} : \mathbb{F}^n \to X, \qquad (x_1, \ldots, x_n) \mapsto \sum_{k=1}^n x_k v_k =: x,$$

is linear and continuous by Lemma 4.2. Since compositions of continuous mappings are again continuous by Corollary 4.6, the claim follows since $X \simeq (\mathbb{F}^n, \| \cdot \|) \simeq Y$ by Theorem 3.8. □

A less obvious example is $c(\mathbb{F}) \simeq c_0(\mathbb{F})$ via the mapping

$$T : c(\mathbb{F}) \to c_0(\mathbb{F}), \qquad (x_1, x_2, x_3, \ldots) \mapsto \left(\lim_{k \to \infty} x_k, x_1 - \lim_{k \to \infty} x_k, x_2 - \lim_{k \to \infty} x_k, \ldots \right).$$

A different generalization of the equivalence of norms is possible for normed vector spaces X, Y with $X \subset Y$. In this case, we call X *continuously embedded* in

Y and write $X \hookrightarrow Y$ if the identity $\mathrm{Id} : X \to Y$ is continuous, i.e., if there exists a $C > 0$ such that

$$\|x\|_Y \le C\|x\|_X \qquad \text{for all } x \in X.$$

For example, $\ell^p(\mathbb{F}) \hookrightarrow \ell^q(\mathbb{F})$ for $1 \le p \le q \le \infty$ and $L^p(\Omega) \hookrightarrow L^q(\Omega)$ for $1 \le q \le p < \infty$. (Note the reverse ordering of p and q!)

Problems

Problem 4.1 *(Operator norm 1 (Lemma 4.3))*
Let $(X, \|\cdot\|_X)$ and $(Y, \|\cdot\|_Y)$ be normed vector spaces and $T \in L(X, Y)$. Show the following relations:

(i) $\displaystyle \|T\|_{L(X,Y)} = \sup_{x \in X, \|x\|_X < 1} \|Tx\|_Y = \sup_{x \in X, \|x\|_X = 1} \|Tx\|_Y = \sup_{x \in X \setminus \{0\}} \frac{\|Tx\|_Y}{\|x\|_X};$

(ii) $\|T\|_{L(X,Y)} = \inf\{C > 0 : \|Tx\|_Y \le C\|x\|_X \text{ for all } x \in X\};$

(iii) $\|Tx\|_Y \le \|T\|_{L(X,Y)}\|x\|_X$ for all $x \in X$.

Problem 4.2 *(Operator norm 2)*
Let X and Y be normed vector spaces and $T \in L(X, Y)$. Show or give a counterexample for the existence of an $x \in X$ with $\|Tx\|_Y = \|T\|_{L(X,Y)}\|x\|_X$.

Problem 4.3 *(Range of operators)*
Let $a, b \in \mathbb{R}$ with $a < b$ and define the mapping $T : C([a, b]) \to C([a, b])$ via

$$[Tx](t) = tx(t) \quad \text{for all } t \in [a, b].$$

(i) Show that T is a continuous linear operator.
(ii) Calculate $\|T\|_{L(C([a,b]),C([a,b]))}$.
(iii) Show or give a counterexample for the closedness of ran T.

Problem 4.4 *(Isomorphisms and separability)*
Let X be a separable normed vector space and $Y \cong X$. Show that Y is separable as well.

Problem 4.5 *(Isomorphisms)*
Show that

$$T : c(\mathbb{F}) \to c_0(\mathbb{F}), \qquad (x_1, x_2, x_3, \dots) \mapsto \left(\lim_{k \to \infty} x_k, x_1 - \lim_{k \to \infty}, x_2 - \lim_{k \to \infty} x_k, \dots \right),$$

is continuously invertible, i.e., that $c_0(\mathbb{F})$ is isomorphic to $c(\mathbb{F})$.

Problem 4.6 *(Closure and preimages of convex sets)*

Let X and Y be normed vector spaces, $A \subset Y$ convex, and $T \in L(X, Y)$. Show that

(i) cl A is convex;

(ii) $T^{-1}(A)$ is convex.

Problem 4.7 *(Embedding of the ℓ^p spaces)*

Show that $\ell^p(\mathbb{R}) \hookrightarrow \ell^q(\mathbb{R})$ for all $1 \leq p \leq q \leq \infty$.

Hint: Consider first the case $\|x\|_p = 1$.

The Uniform Boundedness Principle

5

We now come to one of the core principles of functional analysis: The completeness of Banach spaces implies that certain *pointwise* properties of linear operators in fact hold *uniformly*. In this chapter, we will derive from this some of the fundamental theorems on linear operators; further important consequences will show up in Part III.

Since completeness is a metric property, all these theorems are based on an abstract property of complete metric spaces known as *Baire's theorem*.[1] There are several equivalent versions, of which we require the following.

Theorem 5.1 (Baire)

Let X be a complete metric space and $\{A_n\}_{n \in \mathbb{N}}$ a sequence of closed subsets $A_n \subset X$. If $A := \bigcup_{n \in \mathbb{N}} A_n$ contains an interior point, then there exists a $j \in \mathbb{N}$ such that A_j contains an interior point as well.

Proof. We argue by contradiction. Assume that A contains an interior point, but none of the A_n do. The former means that A contains an open ball $U_0 := U_{\varepsilon_0}(x_0)$; the latter implies that $(X \setminus A_n) \cap U_\varepsilon(x) \neq \emptyset$ for all $n \in \mathbb{N}$, $\varepsilon > 0$, and $x \in X$ (since otherwise A_n would contain an open ball and hence an interior point).

We now define inductively a sequence $\{B_{\varepsilon_n}(x_n)\}_{n \in \mathbb{N}}$ of nested closed balls with $\varepsilon_n < \frac{1}{n}$ as follows: For $n = 1$, we choose $\varepsilon_1 < \min\{1, \varepsilon_0/2\}$ and $x_1 := x_0$. If x_n and ε_n are now

---

[1] This result is also known as the *Baire category theorem*; however, this name stems from historical terminology that is not relevant in our context.

© Springer Nature Switzerland AG 2020
C. Clason, *Introduction to Functional Analysis*, Compact Textbooks in Mathematics, https://doi.org/10.1007/978-3-030-52784-6_5

given as specified, we choose $\varepsilon_{n+1} < \frac{1}{n}$ and $x_{n+1} \in X$ such that

$$B_{\varepsilon_{n+1}}(x_{n+1}) \subset (X \setminus A_{n+1}) \cap U_{\varepsilon_n}(x_n).$$

This is possible since the set on the right-hand side is open and nonempty by assumption. This construction then ensures that

$$U_{\varepsilon_{n+1}}(x_{n+1}) \subset B_{\varepsilon_{n+1}}(x_{n+1}) \subset U_{\varepsilon_n}(x_n) \subset B_{\varepsilon_n}(x_n).$$

The sequence $\{x_n\}_{n\in\mathbb{N}}$ thus satisfies $x_m \in U_{\varepsilon_n}(x_n)$ for all $m \geq n$. Since $\varepsilon_n \to 0$, it is therefore a Cauchy sequence, which due to the completeness of X has a limit $x \in X$. Since $x_n \in B_{\varepsilon_k}(x_k)$ for all $k \leq n$, it follows that $x \in B_{\varepsilon_k}(x_k)$ for all $k \in \mathbb{N}$.[2] On the one hand, this implies that

$$x \in \bigcap_{k\in\mathbb{N}} B_{\varepsilon_k}(x_k) \subset \bigcap_{k\in\mathbb{N}} (X \setminus A_k) = X \setminus \left(\bigcup_{k\in\mathbb{N}} A_k \right) = X \setminus A$$

and hence that $x \notin A$. On the other hand, since the balls are nested, we obtain that $x \in B_{\varepsilon_1}(x_1) \subset U_0 \subset A$ and therefore the desired contradiction. $\qquad\square$

Baire's theorem ensures a particular compatibility between the algebraic and topological structures of complete normed vector spaces. To see this, we define for a subset A of a vector space X the *algebraic interior* or *core*

$$\text{core } A := \{x \in A : \text{for all } h \in X \text{ there is a } \delta > 0 \text{ with } x + th \in A \text{ for all } t \in [0, \delta]\}.$$

Intuitively, for $x \in \text{core } A$ it is possible to go in any direction at least a small distance without leaving A. Furthermore, recall that a set A is *convex* if $\lambda x + (1 - \lambda)y \in A$ for all $x, y \in A$ and $\lambda \in [0, 1]$, i.e., if it contains all line segments between two points in A.

Lemma 5.2 (Core–int)
If X is a Banach space and $A \subset X$ is closed and convex, then $\text{core } A = \text{int } A$.

Proof. The inclusion $\text{int } A \subset \text{core } A$ is a direct consequence of the definition of interior point in normed vector spaces: if $x \in A$ is an interior point, then there exists a $U_\varepsilon(x) \subset A$, implying for all $h \in X$ and $t < \varepsilon \|h\|_X^{-1} =: \delta$ that $x + th \in U_\varepsilon(x) \subset A$.

---

[2] This is precisely the reason why we have used *closed* balls for the construction.

Let conversely $x \in$ core A be given; we assume for simplicity that $x = 0$. (The general case then follows from the translation invariance of the definitions of core and int.) By assumption, we can then find for every $h \in X$ a $t > 0$ sufficiently small that $th \in A$, i.e., $h \in t^{-1}A$ for sufficiently small $t > 0$. We can therefore write $X = \bigcup_{n \in \mathbb{N}}(nA)$. Clearly, core $X = X$, and it follows from Baire's theorem (Theorem 5.1) that $\text{int}(nA) \neq \emptyset$ for some $n \in \mathbb{N}$. But this is possible only if int $A \neq \emptyset$.

It remains to show that $0 \in$ int A. To this end, let $x \in A$ be an interior point, i.e., $U_\varepsilon(x) \subset A$ for some $\varepsilon > 0$. Since $0 \in$ core A by assumption, there exists for $h = -x$ a $\delta > 0$ such that $-\delta x \in A$. By the convexity of A, we thus have for all $y \in U_\varepsilon(x) \subset A$ and $t = \frac{1}{1+\delta} < 1$ that

$$ z := t(-\delta x) + (1-t)y = \frac{\delta}{1+\delta}(y-x) \in A. $$

Since $\|y - x\|_X \leq \varepsilon$, this implies that $z \in U_r(0)$ for $r = \frac{\delta\varepsilon}{1+\delta}$. Conversely, all elements of $U_r(0)$ can be written in this way. This shows that $U_r(0) \subset A$, and hence 0 is an interior point of A.   □

This lemma seems unimpressive but is nevertheless of fundamental importance since it establishes the promised link between algebraic (core) and topological (int) properties. In essence, it guarantees that properties that hold pointwise and are preserved by passing to the limit (and forming convex combinations) in fact hold uniformly. (Of course, there is no free lunch: the closedness of A is a nontrivial requirement that often enough is not given.) In functional analysis, this *uniform boundedness principle* is often applied in the following form.

Theorem 5.3 (Banach–Steinhaus)

Let X be a Banach space and Y a normed vector space. If a subset $\mathcal{T} \subset L(X, Y)$ satisfies

$$ \sup_{T \in \mathcal{T}} \|Tx\|_Y < \infty \qquad \text{for all } x \in X, $$

then

$$ \sup_{T \in \mathcal{T}} \|T\|_{L(X,Y)} < \infty. $$

Proof. We apply the core–int lemma (Lemma 5.2) to the set

$$ A := \left\{ x \in X : \sup_{T \in \mathcal{T}} \|Tx\|_Y \leq 1 \right\}. $$

Since $A = \bigcap_{T \in \mathcal{T}} T^{-1}(B_Y)$ and $T \in L(X, Y)$ is continuous, A is closed. The linearity of T^{-1} and the norm properties imply that $T^{-1}(B_Y)$ is convex; hence A is also convex as the intersection of convex sets. Finally, by assumption $\|Tx\|_Y < \infty$ for all $x \in X$ and $T \in \mathcal{T}$; hence the linearity of T and the homogeneity of the norm ensure that for all $h \in X$ there exists a $\delta > 0$ such that $\|T(\delta h)\|_Y \leq 1$ for all $T \in \mathcal{T}$, implying that $\delta h \in A$ and hence $0 \in \text{core } A$. The core–int lemma thus yields that $0 \in \text{int } A$, i.e., there exists an $\varepsilon > 0$ with $U_\varepsilon(0) \subset A$. But this means that $\|x\|_X < \varepsilon$ implies $\|Tx\|_Y \leq 1$. By definition of the operator norm, we thus have

$$\sup_{T \in \mathcal{T}} \|T\|_{L(X,Y)} \leq \frac{1}{\varepsilon} < \infty. \qquad \qquad \square$$

This implies a very useful result: even the *pointwise* limit of a sequence of continuous operators is continuous.

Corollary 5.4

Let X be a Banach space, Y a normed vector space, and $\{T_n\}_{n \in \mathbb{N}} \subset L(X, Y)$. If the limit $Tx := \lim_{n \to \infty} T_n x \in Y$ exists for all $x \in X$, then $x \mapsto Tx$ defines a continuous linear operator $T \in L(X, Y)$.

Proof. The linearity of T follows as in the proof of Theorem 4.5; it remains to show the continuity. The convergence $T_n x \to Tx \in Y$ for all $x \in X$ directly implies that the sequence $\{T_n x\}_{n \in \mathbb{N}}$ is bounded in Y, i.e., that $\sup_{n \in \mathbb{N}} \|T_n x\|_Y < \infty$. The Banach–Steinhaus theorem (Theorem 5.3) then yields that $\sup_{n \in \mathbb{N}} \|T_n\|_{L(X,Y)} < \infty$, and it follows from the continuity of the norm (Theorem 3.6 (iii)) that

$$\|Tx\|_Y = \lim_{n \to \infty} \|T_n x\|_Y \leq \sup_{n \in \mathbb{N}} \|T_n x\|_Y \leq \sup_{n \in \mathbb{N}} \|T_n\|_{L(X,Y)} \|x\|_X \qquad \text{for all } x \in X.$$

Hence T is bounded and therefore continuous. $\qquad \qquad \square$

The uniform boundedness principle yields three fundamental theorems about continuous linear operators: the open mapping theorem, the bounded inverse theorem, and the closed graph theorem. These three theorems are in fact equivalent: each can be used as a starting point for proving the other two. It is therefore a matter of taste which one is first derived directly from the uniform boundedness principle; one usually starts with the open mapping theorem.

For this, we say that a mapping $f : X \to Y$ between two metric spaces X and Y is *open* if $f(U) \subset Y$ is open for all open sets $U \subset X$. The open mapping theorem now states that a continuous linear operator between Banach spaces is open if and only if it is surjective. In the following, we denote by U_X and U_Y the *open* unit balls in X and Y, respectively.

Theorem 5.5 (Open mapping)

Let X and Y be Banach spaces and $T \in L(X, Y)$. Then the following are equivalent:

(i) T is open;
(ii) there exists a $\delta > 0$ such that $\delta U_Y \subset T(U_X)$;
(iii) T is surjective.

Proof. We show that (i) and (ii) as well as (ii) and (iii) are equivalent.

(i) $\Rightarrow$ (ii) follows directly from the definition of open sets and mappings: since T and U_X are open, $T(U_X)$ is open as well; hence there exists in particular an open ball around $0 = T0 \in T(U_X)$.

(ii) $\Rightarrow$ (i): Let $U \subset X$ be open and $y \in T(U)$ arbitrary. We show that y is an interior point. To this end, choose $x \in U$ with $Tx = y$ as well as $\varepsilon > 0$ with $U_\varepsilon(x) \subset U$. Taking $\delta > 0$ from (ii) and using the linearity of T then yields that

$$U_{\delta\varepsilon}(y) = U_{\delta\varepsilon}(Tx) = Tx + \varepsilon\delta U_Y \subset Tx + \varepsilon T(U_X) = T(U_\varepsilon(x)) \subset T(U),$$

i.e., $y \in \text{int } T(U)$, and hence $T(U)$ is open.

(ii) $\Rightarrow$ (iii) again follows from the linearity of T: for all $y \in Y$, we have $\tilde{y} := \frac{\delta}{2}\|y\|_Y^{-1} y \in \delta U_Y$. By (ii), there thus exists an $\tilde{x} \in U_X$ with $T\tilde{x} = \tilde{y}$, i.e., $x := \frac{2}{\delta}\|y\|_Y \tilde{x}$ satisfies $Tx = y$.

(iii) $\Rightarrow$ (ii): This is the part that requires the uniform boundedness principle. For this, we first note that $A := \text{cl } T(U_X) \subset Y$ is closed and convex (since U_X is convex, so is $T(U_X)$ by linearity of T, and the closure of a convex set is still convex). Furthermore, for all $h \in Y$ there exists an $x \in X$ with $Tx = h$ since T was assumed to be surjective. Setting $\delta := \frac{1}{2}\|x\|_X^{-1} > 0$, we have $\delta x \in U_X$ and hence $\delta h = T(\delta x) \in A$. Since Y is complete, we can apply the core–int lemma (Lemma 5.2) to deduce that $0 \in \text{core } A = \text{int } A$. This implies that the set A contains an open ball around 0 with radius δ for some $\delta > 0$.

We now use the completeness of X to show that $\delta U_Y \subset T(U_X)$ (by reducing δ if necessary). Since we have just proved that $\delta U_Y \subset A = \text{cl } T(U_X)$, we can find for every $y \in Y$ with $\|y\|_Y < \delta$ a sequence $\{x_n\}_{n \in \mathbb{N}} \in U_X$ such that $Tx_n \to y$. However, we cannot yet conclude that $\{x_n\}_{n \in \mathbb{N}}$ itself converges. In order to obtain the desired preimage, we therefore construct a new sequence $\{\tilde{x}_n\}_{n \in \mathbb{N}} \in U_X$ with $\tilde{x}_n \to x \in U_X$ and $Tx = y$. We again proceed inductively. Let first $y \in \delta U_Y$ be given and set $y_0 := y \in \delta U_Y$. Assume now that we have $y_n \in \delta U_Y \subset \text{cl } T(U_X)$. We can then find for $\varepsilon := \frac{\delta}{2} > 0$ an $x_n \in U_X$ with $\|y_n - Tx_n\|_Y < \varepsilon$ such that

$$y_{n+1} := 2(y_n - Tx_n) \in \delta U_Y \subset A.$$

It follows that

$$2^{-(n+1)} y_{n+1} = 2^{-n} y_n - T(2^{-n} x_n) \qquad \text{for all } n \in \mathbb{N}.$$

If we now define $\tilde{x}_m := \sum_{n=0}^{m} 2^{-n} x_n$ for all $m \in \mathbb{N}$, we can solve the last equation and reduce the telescoping sum to deduce that

$$T\tilde{x}_m = y_0 - 2^{-(m+1)} y_{m+1} \to y_0 = y \qquad \text{as } m \to \infty, \tag{5.1}$$

since $\{y_n\}_{n \in \mathbb{N}} \subset \delta U_Y$ is bounded. Furthermore, $\|x_n\|_X < 1$ implies that

$$\sum_{n=0}^{m} 2^{-n} \|x_n\|_X < 2 \qquad \text{for all } m \in \mathbb{N}.$$

Taking the supremum over all $m \in \mathbb{N}$ thus shows that the series $\sum_{n=0}^{\infty} 2^n x_n$ converges absolutely and, since X is complete, therefore converges to some $x \in X$ by Lemma 3.7. The closedness of the unit ball further implies that $x \in 2B_X$, and (5.1) implies that $Tx = y \in \delta U_Y$. Scaling $\delta \mapsto \delta/4$ thus yields an $x \in \frac{1}{2} B_X \subset U_X$ with $Tx = y$ for arbitrary $y \in \delta U_Y$, i.e., $\delta U_Y \subset T(U_X)$. $\square$

Note that only the last part required continuity of T and completeness of X; every open linear mapping between normed vector spaces is therefore surjective. On the other hand, the last part is a strong statement: if a (sufficiently small) $y \in Y$ has a preimage at all, it must have even a *bounded* preimage. (Since T was not assumed to be injective, this need not be the same preimage.) For linear operators, "sufficiently small" is of course not really a restriction since we can always scale the equation $Tx = y$ appropriately. This yields the following crucial result.

Theorem 5.6 (Bounded inverse)
Let X and Y be Banach spaces and let $T \in L(X, Y)$ be bijective. Then $T^{-1} \in L(Y, X)$.

Proof. If T is bijective, it is in particular surjective and hence open by the open mapping theorem (Theorem 5.5). For every open set $U \subset X$, we thus have that $T(U) = (T^{-1})^{-1}(U)$ is open, i.e., the preimages of open sets under T^{-1} are always open. Hence T^{-1} is continuous by Theorem 1.8. $\square$

If T is not surjective, one would like to have at least continuous invertibility on the range of T. The following useful result specifies when this is possible.

Corollary 5.7
Let X and Y be Banach spaces and let $T \in L(X, Y)$ be injective. Then $T^{-1} : \operatorname{ran} T \to X$ is continuous if and only if $\operatorname{ran} T$ is closed.

Proof. If ran $T \subset Y$ is closed, then ran T is a Banach space by Lemma 3.5, and hence the restricted operator $T : X \to \text{ran} \, X$ has a continuous inverse by Theorem 5.6. Conversely, if T^{-1} is continuous, then T is an isomorphism between X and ran T. Since X is a Banach space, ran T is therefore a Banach space as well and hence closed by Lemma 3.5.   □

Again, closedness is a nontrivial assumption; the fact that ran T need not be closed is in fact one of the fundamental difficulties in treating operators between infinite-dimensional spaces. For example, in this case the solution of the operator equation $Tx = y$ is not stable even for $y \in \text{ran} \, T$; the equation is then called *ill-posed*. Such equations actually occur in medical imaging (e.g., in computerized tomography), in parameter identification, and, more generally, in so-called *inverse problems*. Solving such problems therefore requires special *regularization methods*; see, e.g., [6, 11].

We finally show a result analogous to Lemma 3.5 for linear operators. For this, we endow graph $T = \{(x, Tx) : x \in X\} \subset X \times Y$ with the *product norm* $\|(x, y)\|_{X \times Y} := \|x\|_X + \|y\|_Y$.

> **Theorem 5.8 (Closed graph)**
>
> *Let X and Y be Banach spaces and let $T : X \to Y$ be linear. Then T is continuous if and only if graph T is closed.*

Proof. First, if $\{(x_n, y_n)\}_{n \in \mathbb{N}} \subset \text{graph} \, T$ converges to some (x, y) in $X \times Y$, then in particular $x_n \to x$ and $y_n \to y$. If T is continuous, this implies that $Tx_n \to Tx$. On the other hand, $Tx_n = y_n \to y$ by the definition of the graph. The uniqueness of the limit then yields $Tx = y$, i.e., $(x, y) \in \text{graph} \, T$.

For the converse, we first note that if T is linear, it is straightforward to verify that graph T is a subspace of $X \times Y$. If graph T is closed, it is therefore a Banach space by Lemma 3.5. Hence the *canonical projections*

$$P_X : X \times Y \to X, \quad (x, y) \mapsto x, \qquad P_Y : X \times Y \to Y, \quad (x, y) \mapsto y,$$

are linear and continuous (by choice of the product norm). Furthermore, the restriction of P_X to graph T is bijective and therefore has a continuous inverse $Q := (P_X|_{\text{graph} \, T})^{-1} : X \to$ graph T by Theorem 5.6 such that $Qx = (x, Tx)$. It follows that for all $x \in X$,

$$Tx = P_Y(x, Tx) = P_Y Qx.$$

Hence $T = P_Y \circ Q$ is continuous by Corollary 4.6.   □

To show that a linear mapping between Banach spaces is continuous, it thus suffices to show that $x_n \to x$ and $Tx_n \to y$ imply that $Tx = y$. In other words, one can already assume the convergence of $\{Tx_n\}_{n \in \mathbb{N}}$ to *some* limit, which one would have to show as part of the verification of the definition of continuity. For general (nonlinear) mappings, however, this is a strictly weaker property, which nevertheless in many cases is an adequate substitute for continuity. Such mappings are called *closed*.

Problems

Problem 5.1 *(Counterexamples for the core–int Lemma)*
Show in each case by a counterexample that Lemma 5.2 no longer holds if any of the assumptions are dropped:

(i) A is convex;
(ii) A is closed;
(iii) X is complete.

Hint: For (iii), consider $X = c_e(\mathbb{R})$ *together with the set*

$$A = \left\{ x \in c_e(\mathbb{R}) : x_k \leq k^{-1} \text{ for all } k \in \mathbb{N} \right\}.$$

Problem 5.2 *(Banach–Steinhaus without Banach?)*
Show by a counterexample that Theorem 5.3 no longer holds if X is not a Banach space.

Problem 5.3 *(Quadrature formulas as continuous operators)*
We consider integration of continuous functions as a linear operator

$$Q : C([a, b]) \to \mathbb{R}, \qquad f \mapsto \int_a^b f(x)\,dx.$$

The numerical evaluation of Q can be performed via approximation by a sequence of *quadrature formulas*

$$Q_n : C([a, b]) \to \mathbb{R}, \qquad f \mapsto \sum_{i=0}^n w_i^{(n)} f\left(x_i^{(n)}\right),$$

where the $x_i^{(n)} \in [a, b]$ are called *quadrature nodes* and the $w_i^{(n)} \in \mathbb{R}$ *quadrature weights*.

(i) Show that Q_n converges pointwise to Q if and only if
  (a) $Q_n(\varphi) \to Q(\varphi)$ for all φ from a dense subset of $C[a, b]$,
  (b) $\sup_{n \in \mathbb{N}} \sum_{j=1}^n |w_j^{(n)}| < \infty$.
(ii) Show that for the pointwise convergence $Q_n \to Q$, it is sufficient that
  (a) $Q_n(\varphi) \to Q(\varphi)$ for all φ from a dense subset of $C[a, b]$,
  (b) $Q_n(1) \to Q(1)$,
  (c) $w_i^{(n)} \geq 0$ for all $n \in \mathbb{N}$ and $i \in \mathbb{N} \cup \{0\}$.

Problem 5.4 *(Open mappings)*

(i) Show that

$$T : \mathbb{R}^2 \to \mathbb{R}, \qquad (x_1, x_2) \mapsto x_1,$$

is open.

(ii) Is

$$T : \mathbb{R}^2 \to \mathbb{R}^2, \qquad (x_1, x_2) \mapsto (x_1, 0),$$

open?

(iii) Show that open mappings do not necessarily map closed sets to closed sets.

Problem 5.5 *(Method of successive approximation)*

Let X be a Banach space and $A \in L(X, X)$ with $\|A\|_{L(X,X)} < 1$. Show that for all $f \in X$ and arbitrary $\varphi_0 \in X$, the sequence

$$\varphi_n := A\varphi_{n-1} + f, \quad n \in \mathbb{N},$$

converges to the unique solution φ of

$$\varphi - A\varphi = f.$$

Problem 5.6 *(Bilinear forms)*

Let X and Y be Banach spaces and let $B : X \times Y \to \mathbb{F}$ be bilinear and partially continuous, i.e., let $x \mapsto B(x, y)$ be linear and continuous for all $y \in Y$ and let $y \mapsto B(x, y)$ be linear and continuous for all $x \in x$. Show that B is continuous on $X \times Y$.

Quotient Spaces

The fundamental theorem of calculus implies that every continuous function has an antiderivative that is unique up to addition of a constant. Correspondingly, the linear derivative operator is injective "up to a constant". If one is not really interested in this constant, i.e., one wants to make statements that hold uniformly for all constants, then the following formalization of "up to a constant" can be helpful. (We will also need it for the proof of another fundamental theorem, the closed range theorem, Theorem 9.10.)

We start with a general construction[1] and define a *seminorm* on a vector space X as a mapping $|\cdot| : X \to \mathbb{R}^+$ that satisfies the properties (ii) and (iii) in Definition 3.1 (i.e., we allow $|x| = 0$ for $x \neq 0$). It is straightforward to verify that $x \sim y$ if and only if $|x - y| = 0$ defines an equivalence relation. The corresponding equivalence classes

$$[x] := \{y \in X : x \sim y\}$$

form a *quotient space*

$$X/\!\!\sim \; := \{[x] : x \in X\},$$

which is a vector space if addition and scalar multiplication are defined via

$$[x] + [y] := [x + y], \qquad \lambda[x] := [\lambda x], \qquad x, y \in X, \; \lambda \in \mathbb{F}.$$

---

[1] Which is also used for the construction of the $L^p(\Omega)$ spaces.

© Springer Nature Switzerland AG 2020
C. Clason, *Introduction to Functional Analysis*, Compact Textbooks
in Mathematics, https://doi.org/10.1007/978-3-030-52784-6_6

We endow this vector space with the *quotient norm*

$$\|[x]\|_\sim := |x|, \qquad [x] \in X/\sim.$$

Theorem 6.1

The pair $(X/\sim, \|\cdot\|_\sim)$ is a normed vector space. If (X, d) with $d(x, y) := |x - y|$ is a complete metric space, then $(X/\sim, \|\cdot\|_\sim)$ is a Banach space.

Proof. We first show that the quotient norm is a well-defined mapping. For this, take any $x, y \in X$ with $[x] = [y]$, i.e., $|x - y| = 0$. The triangle inequality, Definition 3.1 (iii), then implies that

$$|x| \le |x - y| + |y| = |y| \le |y - x| + |x| = |x|,$$

i.e., $\|[x]\|_\sim = |x| = |y| = \|[y]\|_\sim$.

Homogeneity and the triangle inequality for $\|\cdot\|_\sim$ follow from the corresponding properties of the seminorm. For the nondegeneracy, assume that $\|[x]\|_\sim = 0$; then by definition $|x - 0| = |x| = 0$ and hence $[x] = [0]$. Therefore $(X/\sim, \|\cdot\|_\sim)$ is a normed vector space.

We now show the completeness. Let $\{[x]_n\}_{n \in \mathbb{N}} \subset X/\sim$ be a Cauchy sequence and choose $x_n \in X$ with $[x_n] = [x]_n$ for all $n \in \mathbb{N}$. By definition of the quotient norm and the operations on equivalence classes, $\{x_n\}_{n \in \mathbb{N}} \subset X$ is also a Cauchy sequence in (X, d) and thus converges by assumption to some $x \in X$ in (X, d). It follows that

$$\|[x]_n - [x]\|_\sim = \|[x_n - x]\|_\sim = |x_n - x| \to 0$$

and hence that $[x]_n \to [x]$ in $X/\sim$. $\square$

Another point of view is the following: it follows directly from the seminorm properties that $U := \{x \in X : |x| = 0\}$ is a subspace of X; hence by definition $x \sim y$ if and only if $x - y \in U$. (One could say that the corresponding quotient space is obtained by "factorization" of X through U.) Conversely, one can use this to generate quotient spaces from arbitrary subspaces. Let X be a normed vector space, $U \subset X$ a subspace, and define the *distance* of $x \in X$ to U as

$$d_U(x) := \inf_{u \in U} \|x - u\|.$$

We now set $|x| := d_U(x)$ and consider the corresponding quotient space

$$X/U := \{[x] : x \in X\}, \qquad [x] := \{y \in X : x - y \in U\}$$

and endow it with the norm $\|[x]\|_U := d_U(x)$.

> **Theorem 6.2**
>
> *If X is a normed vector space and $U \subset X$ is a subspace, then $\| \cdot \|_U$ is a seminorm on X/U. If U is closed, then $(X/U, \| \cdot \|_U)$ is also a normed vector space. If in addition X is a Banach space, then $(X/U, \| \cdot \|_U)$ is also a Banach space.*

Proof. First, $\| \cdot \|_U$ is well-defined: $[x] = [y]$ implies that $x - y \in U$ and hence that $y = x - v$ for some $v \in U$. It follows that

$$d_U(y) = d_U(x - v) = \inf_{u \in U} \|x - (v + u)\| = \inf_{\tilde{u} \in U} \|x - \tilde{u}\| = d_U(x),$$

since every $\tilde{u} \in U$ can be written as $\tilde{u} := v + u$ for $u \in U$.

The same argument for $\tilde{u} := \lambda u$ also yields the homogeneity. For the triangle inequality, we use that by the properties of the infimum, for every $x \in X$ and every $\varepsilon > 0$ there exists a $u_\varepsilon \in U$ such that

$$\|x - u_\varepsilon\| \leq \inf_{u \in U} \|x - u\| + \varepsilon = d_U(x) + \varepsilon.$$

We now use this to choose for arbitrary $x, y \in X$ and $\varepsilon > 0$ a corresponding u_ε and v_ε, respectively. Since U is a subspace, $u_\varepsilon + v_\varepsilon \in U$ and therefore

$$d_U(x + y) = \inf_{u \in U} \|(x + y) - u\| \leq \|(x + y) - (u_\varepsilon + v_\varepsilon)\| \leq \|x - u_\varepsilon\| + \|y - v_\varepsilon\|$$

$$\leq d_U(x) + d_U(y) + 2\varepsilon.$$

Since $\varepsilon > 0$ was arbitrary, it follows that

$$\|[x] + [y]\|_U = d_U(x + y) \leq d_U(x) + d_U(y) = \|[x]\|_U + \|[y]\|_U.$$

Hence $\| \cdot \|_U$ is a seminorm on X/U.

Let now U be closed and $[x] \in X/U$ with $\|[x]\|_U = d_U(x) = \inf_{u \in U} \|x - u\| = 0$. Then there exists a sequence $\{u_n\}_{n \in \mathbb{N}} \subset U$ with $\|u_n - x\| \to 0$, i.e., $u_n \to x$. Since U is closed, we also have $x \in U$ and therefore $[x] = 0$. Hence $\| \cdot \|_U$ is even a norm on X/U.

We now apply Theorem 6.1 by showing that if X is a Banach space, then the metric space (X, d_U) is complete. To this end, let $\{x_n\}_{n \in \mathbb{N}}$ be a Cauchy sequence with respect to d_U. We first construct from this a Cauchy sequence with respect to $\| \cdot \|_X$ as follows: As before, we can find for $x_{n+1} - x_n$ and $\varepsilon := 2^{-n}$ a $u_n \in U$ such that

$$\|x_{n+1} - x_n - u_n\| \leq d_U(x_{n+1} - x_n) + 2^{-n} \qquad \text{for all } n \in \mathbb{N}.$$

Setting $z_n := x_n - \sum_{i=1}^{n} u_i$, we then have for all $m = n + p \geq n \in \mathbb{N}$ that

$$\|z_m - z_n\| = \left\|x_{n+p} - x_n - \sum_{i=n+1}^{p} u_i\right\| \leq \sum_{k=0}^{p-1} \|x_{n+k+1} - x_{n+k} - u_{n+k}\|$$

$$\leq \sum_{k=0}^{p-1} d_U(x_{n+k+1} - x_{n+k}) + \sum_{k=0}^{p-1} 2^{-n-k}.$$

Since $\{x_n\}_{n \in \mathbb{N}}$ is a Cauchy sequence with respect to d_U, the first term can be made arbitrarily small for any $p \in \mathbb{N}$ by taking $n \in \mathbb{N}$ sufficiently large. Similarly, the second term is always bounded by 2^{-n+1} and can thus also be made arbitrarily small. This shows that $\{z_n\}_{n \in \mathbb{N}}$ is a Cauchy sequence in the Banach space X and therefore converges to some $z \in X$. On the other hand, since $\sum_{i=1}^{n} u_i \in U$ for all $n \in \mathbb{N}$, this implies that

$$d_U(x_n - z) = \inf_{u \in U} \|x_n - z - u\| \leq \left\|x_n - z - \sum_{i=1}^{n} u_i\right\| = \|z_n - z\| \to 0,$$

i.e., $\{x_n\}_{n \in \mathbb{N}}$ converges with respect to d_U. Hence (X, d_U) is complete.

Finally, since U is closed, we have that $U = \mathrm{cl}\, U = \{x \in X : d_U(x) = 0\}$, and the claim thus follows from Theorem 6.1. □

A special case that will often be useful is $U = \ker T$ for some $T \in L(X, Y)$. Intuitively, this allows us to replace a noninjective operator with an "equivalent" injective operator.

Lemma 6.3
Let X and Y be normed vector spaces, $T \in L(X, Y)$, and $U \subset \ker T \subset X$ a closed subspace. Then there exists a unique $S \in L(X/U, Y)$ such that

(i) $S[x] = Tx$ for all $x \in X$;
(ii) $\|S\|_{L(X/U,Y)} = \|T\|_{L(X,Y)}$.

If $U = \ker T$, then S is injective.

Proof. We set

$$S : X/U \to Y, \qquad [x] \mapsto Tx.$$

The operator S is well-defined, since $y \in [x]$ implies that $y - x \in U \subset \ker T$, i.e., $Ty - Tx = T(y - x) = 0$. It follows directly from the definition that S is linear and satisfies $Tx = S[x]$

for all $x \in X$. Furthermore, for all $x \in X$ and $y \in [x]$,

$$\|S[x]\|_Y = \|Tx\|_Y = \|Ty\|_Y \leq \|T\|_{L(X,Y)}\|y\|_X.$$

For $y = x - v \in U$, it follows that

$$\|S[x]\|_Y \leq \inf_{y \in [x]} \|T\|_{L(X,Y)}\|y\|_X = \|T\|_{L(X,Y)} \inf_{v \in U} \|x - v\|_X = \|T\|_{L(X,Y)}\|[x]\|_U$$

and hence that $\|S\|_{L(X/U,Y)} \leq \|T\|_{L(X,Y)}$. In particular, $S \in L(X/U, Y)$. Conversely, for all $x \in X$,

$$\|Tx\|_Y = \|S[x]\|_{L(X,Y)} \leq \|S\|_{L(X/U,Y)}\|[x]\|_{X/U} = \|S\|_{L(X/U,Y)} \inf_{u \in U} \|x - u\|_X$$

$$\leq \|S\|_{L(X/U,Y)}\|x\|_X,$$

i.e., $\|T\|_{X,Y} \leq \|S\|_{L(X/U,Y)}$.

If $U = \ker T$ (which is always closed as a null space), then $0 = S[x] = Tx$ implies that $x \in \ker T = U$ and hence by definition of the equivalence class that $[x] = [0]$. This shows that $\ker S = \{[0]\}$ and thus that S is injective.   □

The proof shows in particular that the *quotient mapping*

$$Q : X \to X/U, \qquad x \mapsto [x],$$

is linear and continuous with operator norm $\|Q\|_{L(X,X/U)} = 1$ and satisfies $T = S \circ Q$.

Similarly, we can restore the missing surjectivity of T by restriction to ran T; the *continuous* invertibility of course again requires that the range be closed.

Theorem 6.4

Let X and Y be Banach spaces and let $T \in L(X, Y)$ have closed ran T. Then the operator $S : X/\ker T \to$ ran T defined via $S \circ Q = T$ is an isomorphism, i.e.,

$$X/\ker T \simeq \text{ran } T.$$

Proof. By Lemma 6.3, $S : X/\ker T \to$ ran $T = $ ran S is linear, continuous, and (due to the restricted range) bijective. Since Y is a Banach space and ran T is by assumption a closed subspace, ran T is also a Banach space by Lemma 3.5. The bounded inverse theorem (Theorem 5.6) then yields that S^{-1} is continuous as well, and hence S is an isomorphism. □

Corollary 6.5

Let X and Y be Banach spaces and let $T \in L(X, Y)$ be surjective. Then $X/\ker T \simeq Y$.

Problems

Problem 6.1 *(Sequences in quotient spaces)*
Let X be a normed vector space and U a closed subspace of X. Show that for every convergent sequence $\{z_n\}_{n \in \mathbb{N}} \subset X/U$, there exists a convergent sequence $\{x_n\}_{n \in \mathbb{N}} \subset X$ with $Qx_n = z_n$ for all $n \in \mathbb{N}$.

Problem 6.2 *(Open sets in quotient spaces)*
Let X be a normed vector space and U a closed subspace of X. Show that $V \subset X/U$ is open if and only if $Q^{-1}(V) \subset X$ is open.

Problem 6.3 *(Continuous mappings on quotient spaces)*
Let X be a normed vector space, U a closed subspace of X, and (Y, d) a metric space. Show that $f : X/U \to Y$ is continuous if and only if $f \circ Q : X \to Y$ is continuous.

Problem 6.4 *(Quotients of sequence spaces)*
Let $X = \ell^\infty(\mathbb{F})$ and $U = c_0(\mathbb{F})$. Show that $d_U(x) = \limsup_{k \to \infty} |x_k|$.

Part III

Dual Spaces and Weak Convergence

Linear Functionals and Dual Spaces

7

We have already seen that a fundamental difficulty in handling infinite-dimensional spaces is the fact that norm convergence is not equivalent to componentwise convergence; this means that many useful consequences of the Heine–Borel theorem (Theorem 2.5)—in particular, the Bolzano–Weierstraß theorem (Corollary 2.6)—no longer hold. Since these are too useful to give up without a fight, this part of the book is devoted to generalizing componentwise convergence to infinite-dimensional spaces.

The basic idea is the following: On $\mathbb{F}^n$, the component mapping (or canonical projection) $x = (x_1, \dots, x_n) \mapsto x_k \in \mathbb{F}$ for $1 \leq k \leq n$ is linear (clearly) and continuous (clearly with respect to $\| \cdot \|_1$ and hence with respect to any norm by Theorem 3.8). Correspondingly, for an infinite-dimensional normed vector space X over $\mathbb{F}$ we will consider continuous (this is now an explicit requirement) linear mappings from X to $\mathbb{F}$. The space $X^* := L(X, \mathbb{F})$ of all such mappings is called the *dual space* of X; the elements $x^* \in X^*$ are called *continuous linear functionals*. Since $\mathbb{F}$ is complete, Theorem 4.5 implies that X^* together with the operator norm

$$\|x^*\|_{X^*} = \sup_{x \in B_X} |x^*(x)|$$

is a Banach space. It is customary to write

$$\langle x^*, x \rangle_X := x^*(x)$$

in order to stress that the *duality pairing* $\langle \cdot, \cdot \rangle_X : X^* \times X \to \mathbb{F}$ is bilinear, i.e., satisfies

$$\langle \alpha x_1^* + x_2^*, \beta x_1 + x_2 \rangle_X = \alpha\beta \langle x_1^*, x_1 \rangle_X + \alpha \langle x_1^*, x_2 \rangle_X + \beta \langle x_2^*, x_1 \rangle + \langle x_2^*, x_2 \rangle_X$$

© Springer Nature Switzerland AG 2020
C. Clason, *Introduction to Functional Analysis*, Compact Textbooks
in Mathematics, https://doi.org/10.1007/978-3-030-52784-6_7

for all $x_1^*, x_2^* \in X^*$, $x_1, x_2 \in X$, and $\alpha, \beta \in \mathbb{F}$ (even for $\mathbb{F} = \mathbb{C}$!) We recall
Lemma 4.3 (iii), which in the current notation states that

$$|\langle x^*, x \rangle_X| \leq \|x^*\|_{X^*}\|x\|_X \qquad \text{for all } x^* \in X^*, \ x \in X.$$

Clearly, every vector $x \in \mathbb{F}^n$ is uniquely determined by its components; in this
sense, these continuous linear functionals fully characterize the space $\mathbb{F}^n$. The question is now whether infinite-dimensional spaces are similarly characterized by their
dual space (and in particular, whether the corresponding notion of convergence—
which we will define later in Chap. 11—is meaningful). The answer is not obvious
and requires another fundamental theorem of functional analysis, which will be the
topic of the next chapter.

In the remainder of this chapter, we will study some examples of dual spaces.
This is complicated by the fact that functionals are defined by their action on
elements of X and in general cannot be given explicitly (i.e., without referring
to an $x \in X$). However, some dual spaces admit a more concrete representation,
which usually consists in showing that the space is (isometrically) isomorphic
to some other Banach spaces. For example, it is known from linear algebra
that the algebraic dual space (i.e., the vector space of all linear—not necessarily
continuous—functionals) of $\mathbb{F}^n$ has again dimension n. Since all finite-dimensional
vector spaces of the same dimension are isomorphic by Theorem 4.8, it follows that
$(\mathbb{F}^n)^* \simeq \mathbb{F}^n$ (independent of the norm); we will obtain an isometric isomorphism as
a consequence of the following example.

The situation is naturally more complicated in infinite-dimensional spaces. As a
first example, we again consider the sequence spaces $\ell^p(\mathbb{F})$ and $c_0(\mathbb{F})$.

Theorem 7.1

Let $1 \leq p < \infty$ and $1 < q \leq \infty$ be such that $\frac{1}{p} + \frac{1}{q} = 1$ (with $\frac{1}{\infty} := 0$). Then

$$T : \ell^q(\mathbb{F}) \to \ell^p(\mathbb{F})^*, \qquad \langle Tx, y \rangle_{\ell^p} = \sum_{k=1}^{\infty} x_k y_k, \qquad (7.1)$$

is an isometric isomorphism.
The same mapping yields an isometric isomorphism for $\ell^1(\mathbb{F}) \cong c_0(\mathbb{F})^*$.

Proof. We first show that this T is a continuous linear operator. Let $x \in \ell^q(\mathbb{F})$ and $y \in \ell^p(\mathbb{F})$
be given. For $1 < p < \infty$, the *Hölder inequality* yields that

$$\sum_{k=1}^{N} |x_k||y_k| \leq \left(\sum_{k=1}^{N} |x_k|^q \right)^{\frac{1}{q}} \left(\sum_{k=1}^{N} |y_k|^p \right)^{\frac{1}{p}} \leq \|x\|_q \|y\|_p \qquad \text{for all } N \in \mathbb{N}.$$

For $p = 1$ and $q = \infty$, the "outer" inequality follows directly from the definition of the supremum norm. Hence the series $\sum_{k=1}^{\infty} x_k y_k$ is absolutely convergent, and therefore

$$|\langle Tx, y\rangle_{\ell^p}| \leq \sum_{k=1}^{\infty} |x_k y_k| \leq \|x\|_q \|y\|_p.$$

For fixed $x \in \ell^q(\mathbb{F})$, we thus have $Tx \in L(\ell^p(\mathbb{F}), \mathbb{F}) = \ell^p(\mathbb{F})^*$ with

$$\|Tx\|_{\ell^p(\mathbb{F})^*} \leq \|x\|_q. \tag{7.2}$$

In particular, $\|Tx\|_{\ell^p(\mathbb{F})^*} < \infty$, and hence T is well-defined as an operator from $\ell^q(\mathbb{F})$ to $\ell^p(\mathbb{F})^*$. It follows from the definition (7.1) that T is linear and from (7.2) that T is continuous.

To show the bijectivity of T, we use the *unit vectors* $e_k \in \ell^p(\mathbb{F})$, defined for every $1 \leq p \leq \infty$ as

$$[e_k]_j = \begin{cases} 1 & \text{if } k = j, \\ 0 & \text{otherwise.} \end{cases}$$

It is straightforward to verify that $\|e_k\|_p = 1$ for all $1 \leq p \leq \infty$ (which explains the name). If now $Tx = 0 \in \ell^p(\mathbb{F})^*$, then $0 = \langle Tx, e_k\rangle_{\ell^p} = x_k$ for all $n \in \mathbb{N}$ and hence $x = 0$. This shows that T is injective.

For the surjectivity, let $y^* \in \ell^p(\mathbb{F})^*$ be given; we now have to find an $x \in \ell^q(\mathbb{F})$ with $Tx = y^*$. We proceed step by step, comparing the actions of Tx and y^* on certain vectors. First, we have for the unit vectors that $\langle Tx, e_k\rangle_{\ell^p} = x_k$ and hence that $x_k = \langle y^*, e_k\rangle_{\ell^p}$ for all $k \in \mathbb{N}$. This already yields a unique candidate sequence $x = \{x_k\}_{k \in \mathbb{N}}$; it remains to show that $x \in \ell^q(\mathbb{F})$. To this end, consider a finite sequence $y \in c_e(\mathbb{F})$, i.e., $y = \sum_{k=1}^{N} y_k e_k$ for some $N \in \mathbb{N}$ and $y_k \in \mathbb{F}$, $1 \leq k \leq N$. By linearity of y^*, we then have

$$\langle y^*, y\rangle_{\ell^p} = \langle y^*, \sum_{k=1}^{N} y_k e_k\rangle_{\ell^p} = \sum_{k=1}^{N} y_k \langle y^*, e_k\rangle_{\ell^p} = \sum_{k=1}^{N} y_k x_k. \tag{7.3}$$

For $p = 1$ and $q = \infty$, it follows from $x_k = \langle y^*, e_k\rangle_{\ell^1}$ that

$$\|x\|_\infty = \sup_{k \in \mathbb{N}} |x_k| = \sup_{k \in \mathbb{N}} |\langle y^*, e_k\rangle_{\ell^1}| \leq \sup_{y \in B_{\ell^1}} |\langle y^*, y\rangle_{\ell^1}| = \|y^*\|_{\ell^1(\mathbb{F})^*} \tag{7.4}$$

since $\|e_k\|_1 = 1$, which yields $x \in \ell^\infty(\mathbb{F})$. For $1 < p < \infty$, we choose the specific sequence $y \in c_e(\mathbb{F})$ given by

$$y_k := \begin{cases} |x_k|^{q-1} \sigma_k & \text{for } k \leq N, \\ 0 & \text{otherwise,} \end{cases}$$

with $\sigma_k x_k = |x_k|$ and $|\sigma_k| = 1$. For all $1 \le k \le N$, we thus have

$$x_k y_k = |x_k|^q = |x_k|^{p(q-1)} = |y_k|^p,$$

which implies that

$$\sum_{k=1}^{N} |x_k|^q = \sum_{k=1}^{N} x_k y_k = \langle y^*, y \rangle_{\ell^p} \le \|y^*\|_{\ell^p(\mathbb{F})^*} \|y\|_p = \|y^*\|_{\ell^p(\mathbb{F})^*} \left(\sum_{k=1}^{N} |x_k|^q \right)^{\frac{1}{p}}.$$

Dividing by the second term on the right-hand side and using $1 - \frac{1}{p} = \frac{1}{q}$ then yields that

$$\left(\sum_{k=1}^{N} |x_k|^q \right)^{\frac{1}{p}} \le \|y^*\|_{\ell^p(\mathbb{F})^*}.$$

Passing to the limit $N \to \infty$, we now obtain that

$$\|x\|_q \le \|y^*\|_{\ell^p(\mathbb{F})^*}, \tag{7.5}$$

and hence $x \in \ell^q(\mathbb{F})$. This shows that T is surjective and therefore bijective. Since $\ell^p(\mathbb{F})$ is a Banach space for $1 \le p \le \infty$, the inverse of T is even continuous by the bounded inverse theorem (Theorem 5.6).

It remains to show that $\langle Tx, y \rangle_{\ell^p} = \langle y^*, y \rangle_{\ell^p}$ for all $y \in \ell^p(\mathbb{F})$. Comparing (7.3) with (7.1) yields $\langle Tx, y \rangle_{\ell^p} = \langle y^*, y \rangle_{\ell^p}$ for all $y \in c_e(\mathbb{F})$. Since $c_e(\mathbb{F})$ is a dense subspace of ℓ^p for $1 \le p < \infty$ (see Theorem 3.15), Theorem 4.7 implies that $Tx = y^*$ (otherwise T and y^* would be two different extensions of $y^*|_{c_e}$, in contradiction to the uniqueness of the extension). Combining (7.2) with (7.4) or (7.5) thus implies that

$$\|Tx\|_{\ell^p(\mathbb{F})^*} \le \|x\|_q \le \|y^*\|_{\ell^p(\mathbb{F})^*} = \|Tx\|_{\ell^p(\mathbb{F})^*}.$$

Hence T is an isometric isomorphism.

Proceeding in the same way (this time choosing $y_k = \sigma_k$) shows that $\ell^1(\mathbb{F}) \cong c_0(\mathbb{F})^*$. $\quad\square$

The above proof (without passing to the limit) also shows that $(\mathbb{R}^N, \|\cdot\|_p) \cong (\mathbb{R}^N, \|\cdot\|_q)^*$, since in this case $x \in \mathbb{R}^N$ always holds, independently of the norm. On the other hand, the proof does not go through for $\ell^p(\mathbb{F})$ with $p = \infty$, since in this case $c_e(\mathbb{F})$ is *not* dense. In fact, it is possible to show that $\ell^\infty(\mathbb{F})^*$ is strictly larger than $\ell^1(\mathbb{F})$; see Problem 8.6.

A similar construction together with results from measure theory leads to the following representation; the (technical) proof can be found in, e.g., [5, Proposition 13.4].

Theorem 7.2

Let $1 \leq p < \infty$ and $1 < q \leq \infty$ be such that $\frac{1}{p} + \frac{1}{q} = 1$ (with $\frac{1}{\infty} := 0$). Then

$$T : L^q(\Omega) \to L^p(\Omega)^*, \qquad \langle Tf, g \rangle_{L^p} = \int_\Omega f(t)g(t)\,dt,$$

is an isometric isomorphism.

Again, $L^\infty(\Omega)^*$ is strictly larger than $L^1(\Omega)$.[1]

We close this chapter by studying the dual space of quotient spaces. This requires some notation. For a normed vector space X and subsets $A \subset X$ and $B \subset X^*$, we define the *annihilators*

$$A^\perp := \left\{ x^* \in X^* : \langle x^*, x \rangle_X = 0 \quad \text{for all } x \in A \right\},$$

$$B_\perp := \left\{ x \in X : \quad \langle x^*, x \rangle_X = 0 \quad \text{for all } x^* \in B \right\}.$$

These are always closed subspaces of X^* and X, respectively, since for every sequence $\{x_n^*\}_{n \in \mathbb{N}} \subset A^\perp$ with $x_n^* \to x^* \in X^*$ and arbitrary $x \in A$, we have

$$|\langle x^*, x \rangle_X| = |\langle x^* - x_n^*, x \rangle_X| \leq \|x_n^* - x^*\|_{X^*} \|x\|_X \to 0,$$

and hence $x^* \in A^\perp$ (and similarly for $B_\perp$). Furthermore, $X^\perp = \{0\}$ and $\{0\}^\perp = X^*$.
We now show that $(X/U)^* \cong U^\perp$ if U is a closed subspace.

Theorem 7.3

Let $U \subset X$ be a closed subspace. Then

$$T : (X/U)^* \to U^\perp, \qquad \langle Tu^*, x \rangle_X = \langle u^*, [x] \rangle_{X/U} \quad \text{for all } x \in X,$$

is an isometric isomorphism.

Proof. We first show that T is well-defined. For all $u^* \in (X/U)^*$ and $x \in U$,

$$\langle Tu^*, x \rangle_X = \langle u^*, [x] \rangle_{X/U} = \langle u^*, [0] \rangle_{X/U} = 0,$$

and hence $Tu^* \in U^\perp$. The injectivity of T is shown analogously.

---

[1] A related result in measure theory is the Radon–Riesz theorem, which identifies the dual space of $C(\Omega)$ with the space of regular Borel measures; see, e.g., [5, Proposition 13.6].

Let now $x^* \in U^\perp$ be given and define $u^* \in (X/U)^*$ by $\langle u^*, [x] \rangle_{X/U} = \langle x^*, x \rangle_X$ for all $x \in X$. Since $x^* \in U^\perp$, this definition is independent of the choice of the representative $x \in [x]$. Clearly, u^* is a linear functional. To deduce the surjectivity of T, it remains to show that u^* is continuous. For all $[x] \in X/U$, we have

$$|\langle u^*, [x] \rangle_{X/U}| = |\langle x^*, y \rangle_X| \leq \|x^*\|_{X^*} \|y\|_X \qquad \text{for all } y \in [x]$$

and hence

$$|\langle u^*, [x] \rangle_{X/U}| \leq \|x^*\|_{X^*} \inf_{y \in [x]} \|y\|_X = \|x^*\|_{X^*} \inf_{u \in U} \|x - u\|_X = \|x^*\|_X \|[x]\|_{X/U}.$$

This shows that $u^* \in (X/U)^*$ and that $\|u^*\|_{(X/U)^*} \leq \|x^*\|_X = \|Tu^*\|_X$. Hence T is a bijective operator between Banach spaces (since $U^\perp \subset X^*$ is closed) and therefore continuously invertible by Theorem 5.6, i.e., an isomorphism.

On the other hand, we have for all $x \in X$ that

$$|\langle Tu^*, x \rangle_X| = |\langle u^*, [x] \rangle_{X/U}| \leq \|u^*\|_{(X/U)^*} \|[x]\|_{X/U} \leq \|u^*\|_{(X/U)^*} \|x\|_X$$

since the quotient mapping $x \mapsto [x]$ has operator norm 1. It follows from the definition of the operator norm that $\|u^*\|_{(X/U)^*} = \|Tu^*\|_X$, i.e., T is an isometry. $\square$

Problems

Problem 7.1 *(Norms of functionals)*
Determine the operator norm of the following linear functionals:

(i) $x^* \in \ell^1(\mathbb{R})^*$ with $\langle x^*, x \rangle_1 := \sum_{k=1}^\infty (1 - \frac{1}{k}) x_k$ for all $x \in \ell^1(\mathbb{R})$;
(ii) $x^* \in c_0(\mathbb{R})^*$ with $\langle x^*, x \rangle_{c_0} := \sum_{k=1}^\infty \frac{1}{2^{k-1}} x_k$ for all $x \in c_0(\mathbb{R})$.

In each case, is the supremum in the definition of the operator norm attained?

Problem 7.2 *(Null space of unbounded functionals)*
Let X be a normed vector space and $f : X \to \mathbb{F}$ an unbounded linear functional, i.e., there exists no $C > 0$ such that $|f(x)| \leq C \|x\|_X$ for all $x \in X$. Show that $\ker f$ is dense in X, but $\ker f \neq X$.

Problem 7.3 *(Dual space of $c_0(\mathbb{F})$)*
Show that

$$T : \ell^1(\mathbb{F}) \to c_0(\mathbb{F})^*, \qquad \langle Tx, y \rangle_{c_0} = \sum_{k=1}^\infty x_k y_k,$$

is an isometric isomorphism, i.e., $c_0(\mathbb{F})^* \cong \ell^1(\mathbb{F})$.

Problem 7.4 *(Dual space of $c(\mathbb{F})$)*

(i) Show that

$$T : \mathbb{F} \times \ell^1(\mathbb{F}) \to c(\mathbb{F})^*, \qquad \langle T(a, x), y \rangle_c = a \lim_{k \to \infty} y_n + \sum_{k=1}^{\infty} x_k y_k$$

is an isometric isomorphism.

Hint: Choose a suitable norm on $\mathbb{F} \times \ell^1(\mathbb{F})$.

(ii) Deduce that the dual spaces of $c(\mathbb{F})$ and of $c_0(\mathbb{F})$ are isometrically isomorphic.

The Hahn–Banach Theorem

8

We now come to a second fundamental principle of functional analysis connecting algebraic and topological concepts: linearity and continuity are compatible in the sense that a functional defined on a subspace can be extended to the whole space in such a way that both linearity and boundedness are preserved. Like the uniform boundedness principle, this principle is also based on an abstract property, this time of real-valued sublinear functions (or, as we will see later, of convex sets).

Theorem 8.1 (Hahn–Banach)

Let X be a vector space over $\mathbb{R}$ and let $p : X \to \mathbb{R}$ be sublinear, i.e.,

(i) $p(\lambda x) = \lambda p(x)$ for all $x \in X$ and $\lambda \geq 0 \in \mathbb{R}$;
(ii) $p(x + y) \leq p(x) + p(y)$ for all $x, y \in X$.

Let furthermore $U \subset X$ be a subspace and let $f_0 : U \to \mathbb{R}$ be linear with $f_0(x) \leq p(x)$ for all $x \in U$. Then there exists a linear extension $f : X \to \mathbb{R}$ with

(i) $f(x) = f_0(x)$ for all $x \in U$;
(ii) $f(x) \leq p(x)$ for all $x \in X$.

Proof. The proof proceeds in two steps. We first show inductively that f_0 can be extended in the specified manner to subspaces of increasing dimension. In the second step, we show that this process terminates even for infinite-dimensional spaces.

Let U and f_0 be given as above. We then choose $x_1 \in X \setminus U$ and extend f_0 from U to

$$U_1 := \{x + \lambda x_1 : x \in U, \lambda \in \mathbb{R}\}$$

© Springer Nature Switzerland AG 2020
C. Clason, *Introduction to Functional Analysis*, Compact Textbooks
in Mathematics, https://doi.org/10.1007/978-3-030-52784-6_8

by setting $f_1(x + \lambda x_1) := f_0(x) + \lambda \alpha$ for some $\alpha \in \mathbb{R}$. This definition ensures that f_1 is still linear; it remains to choose α such that $f_1 \le p$. First, the linearity of f_0 and the sublinearity of p imply for all $x, y \in U$ that

$$f_0(x) + f_0(y) = f_0(x + y) \le p(x + y) \le p(x - x_1) + p(x_1 + y),$$

which after rearranging yields

$$f_0(x) - p(x - x_1) \le p(x_1 + y) - f_0(y) \qquad \text{for all } x, y \in U.$$

This inequality remains valid if we take the supremum over all $x \in U$ and the infimum over all $y \in U$. We can thus find an $\alpha \in \mathbb{R}$ such that

$$\sup_{x \in U} f_0(x) - p(x - x_1) \le \alpha \le \inf_{y \in U} p(x_1 + y) - f_0(y). \qquad (8.1)$$

We next show that for this choice of α, we have $f_1(y) \le p(y)$ for all $y \in U_1$, i.e., $y = x + \lambda x_1$ for some $x \in U$ and $\lambda \in \mathbb{R}$. If $\lambda > 0$, we can estimate the infimum in the second inequality of (8.1) from above by $x \in U$ to obtain

$$f_0(x) + \alpha \le p(x_1 + x) \qquad \text{for all } x \in U.$$

Since p is sublinear and $\lambda > 0$, it follows that

$$f_1(x + \lambda x_1) = f_0(x) + \lambda \alpha = \lambda \left(f_0 \left(\tfrac{x}{\lambda} \right) + \alpha \right) \le \lambda p \left(x_1 + \tfrac{x}{\lambda} \right) = p(x + \lambda x_1).$$

If $\lambda < 0$, we can instead estimate the supremum in the first inequality of (8.1) from below to obtain

$$f_0(x) - \alpha \le p(x - x_1) \qquad \text{for all } x \in U.$$

Since $-\lambda > 0$, it now follows that

$$f_1(x + \lambda x_1) = -\lambda \left(f_0 \left(\tfrac{x}{-\lambda} \right) - \alpha \right) \le (-\lambda) p \left(\tfrac{x}{-\lambda} - x_1 \right) = p(x + \lambda x_1).$$

If $\lambda = 0$, we have $f_1(x) = f_0(x) \le p(x)$. Hence $f_1(y) \le p(y)$ for all $y \in U_1$.

If X is finite-dimensional, we can continue this process until $U_n = X$; for separable spaces we use induction. The general case, however, requires heavy machinery: *Zorn's lemma*, which is equivalent to the axiom of choice (and to the well-ordering principle) and guarantees that a nonempty partially ordered set for which every totally ordered subset is bounded from above contains a maximal element.[1] To apply it, we first define the set of all linear extensions satisfying the desired properties,

$$\mathcal{A} := \{(W, f) : W \supset U, \ f : W \to \mathbb{R} \ \text{ with } \ f|_U = f_0, \ f \leq p\},$$

and endow $\mathcal{A}$ with the partial order

$$(W_1, f_1) \preceq (W_2, f_2) \qquad \text{if and only if} \qquad W_1 \subset W_2, \quad f_2|_{W_1} = f_1.$$

Clearly, $(U, f_0) \in \mathcal{A}$, and hence $\mathcal{A} \neq \emptyset$. Let now $\mathcal{B} \subset \mathcal{A}$ be a totally ordered subset, i.e., $a \preceq b$ or $b \preceq a$ holds for all $a, b \in \mathcal{B}$. We set

$$W_* := \bigcup_{(W, f) \in \mathcal{B}} W$$

as well as

$$f_* : W_* \to \mathbb{R}, \qquad f_*(x) = f(x) \quad \text{for all } x \in W, \ (W, f) \in \mathcal{B}.$$

Since $\mathcal{B}$ is totally ordered with respect to linear extension, this definition is not vacuous (which is not obvious). Furthermore, W_* is a subspace and f_* is linear. Then by construction, $(W_*, f_*) \succeq (W, f)$ for all $(W, f) \in \mathcal{B}$, and hence (W_*, f_*) is an upper bound. Zorn's lemma thus ensures that there exists a maximal element (U_*, f). This implies that $U_* = X$, since otherwise we could use the above construction to obtain a further extension of f to $U_1 \supsetneq U_*$, in contradiction to the maximality of (U_*, f). □

Note that this linear extension is not unique in general: different choices of α satisfying (8.1) lead to different extensions.

The use of this result for functional analysis rests on the fact that norms are convex and therefore sublinear.[2] We can therefore define a continuous linear functional by specifying its action on a subspace and then extending it to the whole space. This guarantees that the dual space contains enough elements for our purpose.[3]

---

[1] An admittedly unwieldy statement, which led to the following famous quotation (attributed to Jerry L. Bona): "The Axiom of Choice is obviously true, the well–ordering theorem is obviously false; and who can tell about Zorn's lemma?"

[2] Since the convexity of a norm is equivalent to the convexity of the corresponding unit ball, it is possible to extend the Hahn–Banach theorem to *locally convex vector spaces*; these are topological vector spaces where— intuitively—the neighborhoods defined by the topology are convex. Such spaces have many of the properties shown in the following; see, e.g., [7, Chapter IV] or [26].

[3] In fact, [17] shows that the Hahn–Banach theorem is equivalent to the statement that for a Banach space, $X \neq \{0\}$ implies that $X^* \neq \{0\}$ as well.

Theorem 8.2 (Hahn–Banach, extension)
Let X be a normed vector space over $\mathbb{F}$, $U \subset X$ a subspace, and $u^* \in U^*$. Then there exists an $x^* \in X^*$ with $\langle x^*, x \rangle_X = \langle u^*, x \rangle_U$ for all $x \in U$ and $\|x^*\|_{X^*} = \|u^*\|_{U^*}$.

Proof. We first consider the case $\mathbb{F} = \mathbb{R}$ and define, for given $u^* \in U^*$,

$$p : X \to \mathbb{R}^+, \qquad p(x) = \|u^*\|_{U^*} \|x\|_X.$$

Then p is sublinear and satisfies

$$\langle u^*, x \rangle_U \leq \|u^*\|_{U^*} \|x\|_X = p(x) \qquad \text{for all } x \in U.$$

The Hahn–Banach theorem (Theorem 8.1) then yields a linear extension $x^* : X \to \mathbb{R}$ with $\langle x^*, x \rangle_X \leq p(x)$ for all $x \in X$. It remains to show that x^* is continuous and has the claimed norm. For this, we use the linearity of x^*: for all $x \in X$,

$$-\langle x^*, x \rangle_X = \langle x^*, -x \rangle_X \leq p(-x) = \|u^*\|_{U^*} \|-x\|_X = \|u^*\|_{U^*} \|x\|_X.$$

Together with $\langle x^*, x \rangle_X \leq p(x)$ for all $x \in X$, this implies that

$$|\langle x^*, x \rangle_X| \leq \|u^*\|_{U^*} \|x\|_X \qquad \text{for all } x \in X,$$

i.e., x^* is continuous and satisfies $\|x^*\|_{X^*} \leq \|u^*\|_{U^*}$. Since x^* is an extension, we obtain that

$$\|u^*\|_{U^*} = \sup_{x \in U \setminus \{0\}} \frac{|\langle u^*, x \rangle_U|}{\|x\|_X} = \sup_{x \in U \setminus \{0\}} \frac{|\langle x^*, x \rangle_X|}{\|x\|_X} \leq \sup_{x \in X \setminus \{0\}} \frac{|\langle x^*, x \rangle_X|}{\|x\|_X} = \|x^*\|_{X^*}$$

and hence that $\|x^*\|_{X^*} = \|u^*\|_{U^*}$.

The case $\mathbb{F} = \mathbb{C}$ is reduced to the first case. Every vector space over $\mathbb{C}$ can also be considered a vector space over $\mathbb{R}$ (by considering only scalars from $\mathbb{R}$); we denote the real normed vector spaces corresponding to X and U by $X_{\mathbb{R}}$ and $U_{\mathbb{R}}$, respectively (endowed with the same norm). For given $u^* \in X^*$ we then take the real part $u_{\mathbb{R}}^* := \operatorname{Re} u^*$, defined by $\langle u_{\mathbb{R}}^*, x \rangle_{U_{\mathbb{R}}} := \operatorname{Re} \langle u^*, x \rangle_U$. Then $u_{\mathbb{R}}^* \in (U_{\mathbb{R}})^*$ and $\|u_{\mathbb{R}}^*\|_{(U_{\mathbb{R}})^*} \leq \|u^*\|_{U^*}$ as well as (by $\mathbb{C}$-linearity and $\operatorname{Im}(x) = -\operatorname{Re}(ix)$)

$$\langle u^*, x \rangle_U = \operatorname{Re} \langle u^*, x \rangle_U + i \operatorname{Im} \langle u^*, x \rangle_U = \langle u_{\mathbb{R}}^*, x \rangle_{U_{\mathbb{R}}} - i \langle u_{\mathbb{R}}^*, ix \rangle_{U_{\mathbb{R}}} \qquad \text{for all } x \in U.$$

Let now $x_{\mathbb{R}}^*$ be the ($\mathbb{R}$-)linear extension of $u_{\mathbb{R}}^*$ constructed as above and set

$$\langle x^*, x \rangle_X := \langle x_{\mathbb{R}}^*, x \rangle_{X_{\mathbb{R}}} - i \langle x_{\mathbb{R}}^*, ix \rangle_{X_{\mathbb{R}}} \qquad \text{for all } x \in X.$$

This defines a $\mathbb{C}$-linear extension $x^* : X \to \mathbb{C}$. To show that it is continuous, for given $x \in X$ choose $\sigma \in \mathbb{C}$ with $|\sigma| = 1$ such that $|\langle x^*, x \rangle_X| = \sigma \langle x^*, x \rangle_X \in \mathbb{R}$. Then

$$|\langle x^*, x \rangle_X| = \sigma \langle x^*, x \rangle_X = \langle x^*, \sigma x \rangle_X = \langle x^*_{\mathbb{R}}, \sigma x \rangle_{X_{\mathbb{R}}} \leq \|x^*_{\mathbb{R}}\|_{(X_{\mathbb{R}})^*} \|x\|_X,$$

which implies that $\|x^*\|_{X^*} = \|x^*_{\mathbb{R}}\|_{(X_{\mathbb{R}})^*} = \|u^*_{\mathbb{R}}\|_{(U_{\mathbb{R}})^*} \leq \|u^*\|_{U^*}$. As in the real case, we then show that $\|u^*\|_{U^*} \leq \|x^*\|_{X^*}$ and hence that $\|x^*\|_{X^*} = \|u^*\|_{U^*}$. □

Note that in contrast to Theorem 4.7, neither density nor closedness of U was assumed, but at the cost of the extension not being unique.

We can use the Hahn–Banach extension theorem (Theorem 8.2) to derive a number of very useful results in an elegant manner. The next theorem shows that the dual space X^* is sufficiently "fine" to distinguish the elements of X; this guarantees that a normed vector space is completely characterized by its dual space.

Theorem 8.3

Let X be a normed vector space and $x \in X \setminus \{0\}$. Then there exists a norming functional *$x^* \in X^*$ with $\|x^*\|_{X^*} = 1$ and $\langle x^*, x \rangle_X = \|x\|_X$.*

Proof. Define $U = \{\lambda x : \lambda \in \mathbb{F}\}$ and $u^* : U \to \mathbb{F}$ via

$$\langle u^*, \lambda x \rangle_U = \lambda \|x\|_X \qquad \text{for all } \lambda \in \mathbb{F}.$$

In particular, taking $\lambda = 1$ yields $\langle u^*, x \rangle_U = \|x\|_X$. Furthermore, $|\langle u^*, \lambda x \rangle_X| = \|\lambda x\|_X$ and hence $\|u^*\|_{U^*} = 1$. We thus obtain the norming functional by extending u^* to X using Theorem 8.2. □

The norming functional can be interpreted as a generalization to normed vector spaces of the sign of a real number.

This characterization is illustrated by the following results.

Corollary 8.4

Let X be a normed vector space. Then

$$\|x\|_X = \max_{x^* \in B_{X^*}} |\langle x^*, x \rangle_X| \qquad \text{for all } x \in X.$$

Proof. This follows from

$$|\langle x^*, x \rangle_X| \leq \|x^*\|_{X^*} \|x\|_X \leq \|x\|_X \qquad \text{for all } x^* \in B_{X^*}, \ x \in X,$$

with equality for the norming functional from Theorem 8.3.   □

Note that in contrast to the definition of the operator norm $\|x^*\|_{X^*}$, the supremum is here always attained.

Corollary 8.5

Let X be a normed vector space, $U \subset X$ a closed subspace, and $x_0 \in X \setminus U$. Then there exists an $x^* \in X^*$ such that $\langle x^*, x \rangle_X = 0$ for all $x \in U$ and $\langle x^*, x_0 \rangle_X \neq 0$.

Proof. Since U is a closed subspace, X/U is a Banach space by Theorem 6.1 and hence in particular a normed vector space. Let $Q : X \to X/U$, $x \mapsto [x]$, be the corresponding quotient mapping. Since $x_0 \notin U$, we have that $Qx_0 \neq [0]$, and hence Theorem 8.3 yields a $q^* \in (X/U)^*$ such that $\langle q^*, Qx_0 \rangle_{X/U} = \|Qx_0\|_{X/U} \neq 0$. We now take $x^* := q^* \circ Q \in L(X, \mathbb{F})$, which has the desired properties since $Qx = [0]$ for all $x \in U$.   □

Using the definition of annihilators, we can briefly write $x^* \in U^\perp$ if $\langle x^*, x \rangle_X = 0$ for all $x \in U$.

Corollary 8.6

Let X be a normed vector space and $U \subset X$ a subspace. Then the following are equivalent:

(i) U is dense in X;

(ii) $U^\perp = \{0\}$.

Proof. (i) $\Rightarrow$ (ii): Let $x^* \in U^\perp$, i.e., $\langle x^*, x \rangle_X = 0$ for all $x \in U$. This implies that x^* is a linear extension of the null operator $0 \in L(U, \mathbb{F}) = U^*$ on X. Since U is dense in X, Theorem 4.7 implies that $\|x^*\|_{X^*} = \|x^*\|_{U^*} = 0$ and hence that $x^* = 0$.

(ii) $\Rightarrow$ (i): Assume to the contrary that $\operatorname{cl} U \neq X$. Then there exists an $x_0 \in X \setminus (\operatorname{cl} U)$, and hence Corollary 8.5 yields an $x^* \in X^*$ such that $x^* \neq 0$ and $\langle x^*, x \rangle_X = 0$ for all $x \in U \subset \operatorname{cl} U$, i.e., $0 \neq x^* \in U^\perp$. The claim now follows by contraposition.   □

Corollary 8.7

Let X be a normed vector space. If X^* is separable, then X is separable as well.

Proof. Let $U^* := \{x_n^* : n \in \mathbb{N}\} \subset X^*$ be dense in X^*. By definition of the operator norm, we can then find for every x_n^* an $x_n \in B_X$ with $\langle x_n^*, x_n \rangle_X \geq \frac{1}{2}\|x_n^*\|_{X^*}$. We now use Corollary 8.6 to show that the subspace $U := \mathrm{lin}\{x_1, \ldots, x_n\}$ is dense in X. To this end, take $x^* \in U^\perp$. For all x_n^* and x_n, we then obtain from the definition of the operator norm, the fact that $x_n \in U$, and the reverse triangle inequality that

$$\|x^* - x_n^*\|_{X^*} \geq |\langle x^* - x_n^*, x_n \rangle_X| = |\langle x_n^*, x_n \rangle_X| \geq \frac{1}{2}\|x_n^*\|_{X^*}$$

$$\geq \frac{1}{2}|\|x^*\|_{X^*} - \|x^* - x_n^*\|_{X^*}|. \tag{8.2}$$

Since U^* is dense in X^*, we have $\inf_{x_n^* \in U^*} \|x^* - x_n^*\|_{X^*} = 0$. Taking the infimum over all $n \in \mathbb{N}$ on both sides of (8.2) thus yields $0 \geq \frac{1}{2}\|x^*\|_{X^*}$ and therefore $x^* = 0$. Hence U is dense in X. Since rational linear combinations of the x_n form a countable and dense subset of U, the claim follows. $\qquad\square$

Since by Theorem 3.15, the space $\ell^1(\mathbb{F})$ is separable but $\ell^\infty(\mathbb{F})$ is not, $\ell^1(\mathbb{F})$ cannot be isomorphic to $\ell^\infty(\mathbb{F})^*$. Similarly, $L^1(\Omega)$ is not isomorphic to $L^\infty(\Omega)^*$. However, the Hahn–Banach theorem is not constructive (due to the use of the axiom of choice in the form of Zorn's lemma) and doesn't allow specifying an explicit element of $\ell^\infty(\mathbb{F})^*$ that is not representable in $\ell^1(\mathbb{F})$.[4]

Corollary 8.8

Let X be a normed vector space and $U \subset X$ a subspace. Then $(U^\perp)_\perp = \mathrm{cl}\, U$.

Proof. For all $x \in U$, we have by definition that $\langle x^*, x \rangle_X = 0$ for all $x^* \in U^\perp$ and hence that $x \in (U^\perp)_\perp$. This shows that $U \subset (U^\perp)_\perp$. Let now $x \in \mathrm{cl}\, U \setminus U$ and consider a sequence $\{x_n\}_{n \in \mathbb{N}} \subset U$ with $x_n \to x$. Since annihilators are always closed, it follows that $x \in (U^\perp)_\perp$ and hence that $\mathrm{cl}\, U \subset (U^\perp)_\perp$. Conversely, let $x \notin \mathrm{cl}\, U$. By Corollary 8.5, there then exists an $x^* \in U^\perp$ with $\langle x^*, x \rangle_X \neq 0$, i.e., $x \notin (U^\perp)_\perp$. $\qquad\square$

However, for $U \subset X^*$ we in general have only $\mathrm{cl}\, U \subset (U_\perp)^\perp$.

---

[4]In fact, this is provably impossible, since replacing the axiom of choice—in the weakest form that is required to derive the Hahn–Banach theorem—by a weaker axiom *does* allow proving that $\ell^1(\mathbb{F}) \cong \ell^\infty(\mathbb{F})^*$; see [24, §§ 29.37–38].

As indicated at the beginning of this chapter, the Hahn–Banach theorem can also be interpreted as a statement about convex sets: Corollary 8.5 shows in particular that for all $x, y \in X$ with $x \neq y$, there exists an $x^* \in X^*$ with $\langle x^*, x - y \rangle_X \neq 0$; we say that X^* *separates* the points in X. Intuitively, we can define for $x^* \in X^*$ and $\alpha \in \mathbb{R}$ (in the case $\mathbb{F} = \mathbb{R}$) the *hyperplane*

$$H_\alpha := \left\{ x \in X : \langle x^*, x \rangle_X = \alpha \right\}.$$

This means that for given x and y with $x \neq y$, there always exists a hyperplane that separates X into *half-spaces* H_α^- and H_α^+ such that

$$x \in H_\alpha^- := \left\{ x \in X : \langle x^*, x \rangle_X < \alpha \right\}, \qquad y \in H_\alpha^+ := \left\{ x \in X : \langle x^*, x \rangle_X > \alpha \right\}.$$

We now generalize this to the separation of sets, where again convexity will be crucial.

Theorem 8.9 (Hahn–Banach, separation)

Let X be a normed vector space, $A \subset X$ nonempty, open, and convex, and let $x_0 \in X \setminus A$. Then there exists an $x^* \in X^*$ such that

$$\mathrm{Re}\, \langle x^*, x \rangle_X < \mathrm{Re}\, \langle x^*, x_0 \rangle_X \qquad \textit{for all } x \in A.$$

Proof. We first assume that $\mathbb{F} = \mathbb{R}$ so that we can apply the Hahn–Banach theorem. For this, we need a suitable sublinear functional; this will be the *Minkowski functional* for $A \subset X$,

$$m_A : X \to [0, \infty], \qquad x \mapsto \inf \left\{ t > 0 : \tfrac{1}{t} x \in A \right\}.$$

Intuitively, this functional indicates how far some $x \in X$ has to be "pulled back" toward 0 until it lies in A. (If $A = B_X$, then $m_A(x) = \|x\|_X$, which already illustrates the connection to the extension theorem.) We first have to show that such a t in fact exists, i.e., that the infimum is finite. Assume first that $0 \in A$; the general case will later follow by translation. In this case, we have for all $x \in X$ and all t large enough that $\tfrac{1}{t} x \in A$ since A was assumed to be open (and hence contains a ball of radius $\varepsilon > \tfrac{1}{t} \|x\|_X$). The next step is to show sublinearity. The definition already implies that $m_A(\lambda x) = \lambda m_A(x)$ for all $x \in X$ and $\lambda > 0$. Let now $x, y \in X$ and $\varepsilon > 0$ be arbitrary. By the properties of the infimum, we can thus find $t, s > 0$ such that

$$t \leq m_A(x) + \varepsilon, \quad \tfrac{1}{t} x \in A, \qquad s \leq m_A(y) + \varepsilon, \quad \tfrac{1}{s} y \in A.$$

The convexity of A then implies that

$$\frac{1}{t+s}(x+y) = \frac{t}{t+s}\left(\tfrac{1}{t}x\right) + \frac{s}{t+s}\left(\tfrac{1}{s}y\right) \in A$$

and hence that

$$m_A(x+y) \le t+s \le m_A(x) + m_A(y) + 2\varepsilon.$$

Since $\varepsilon > 0$ was arbitrary, this yields $m_A(x+y) \le m_A(x) + m_A(y)$.

We now show that m_A separates x_0 and A; we then construct from this a *linear* functional on a suitable subspace that we can extend using Theorem 8.1. First, $\frac{1}{t}x_0 \notin A$ for all $t < 1$ (otherwise we would have $x_0 = t\frac{x_0}{t} + (1-t)0 \in A$ for some $t \in (0,1)$, since $0 \in A$ and A is convex) and hence

$$m_A(x_0) \ge 1. \tag{8.3}$$

Conversely, $m_A(x) \ge 1$ implies that $\frac{1}{t}x \in X \setminus A$ for all $t \in (0,1)$. Choosing now $\{t_n\}_{n \in \mathbb{N}} \subset (0,1)$ with $t_n \to 1$, it follows that $x = \lim_{n \to \infty} \frac{1}{t_n}x \in X \setminus A$ since $X \setminus A$ is closed as the complement of an open set. We thus have

$$m_A(x) < 1 \qquad \text{for all } x \in A. \tag{8.4}$$

As in the proof of Theorem 8.3, we now define the subspace $U := \{\lambda x_0 : \lambda \in \mathbb{R}\}$ and a linear functional $u^* : U \to \mathbb{R}$ via

$$\langle u^*, \lambda x_0 \rangle_U = \lambda m_A(x_0) \qquad \text{for all } \lambda \in \mathbb{R}.$$

The sublinearity of m_A together with $m_A \ge 0$ and $m_A(0) = 0$ then implies that

$$\text{for } \lambda > 0: \qquad \langle u^*, \lambda x_0 \rangle_X = \lambda m_A(x_0) = m_A(\lambda x_0),$$

$$\text{for } \lambda \le 0: \qquad \langle u^*, \lambda x_0 \rangle_X = \lambda m_A(x_0) \le 0 \le m_A(\lambda x_0).$$

Hence $u^* \le m_A$ on U. The Hahn–Banach theorem thus yields a linear extension $x^* : X \to \mathbb{R}$ with $x^* \le m_A$ on X. It remains to show that x^* is continuous and still separates x_0 and A.

Since we have assumed that $0 \in A$ and A is open, there exists for sufficiently small $\varepsilon > 0$ a closed ball $B_\varepsilon(0) \subset A$. Hence $\varepsilon \frac{x}{\|x\|_X} \in B_\varepsilon(0) \subset A$ for all $x \in X$, and therefore

$$\langle x^*, x \rangle_X \le m_A(x) \le \frac{1}{\varepsilon}\|x\|_X \qquad \text{for all } x \in X.$$

Arguing analogously for $-x$ yields $x^* \in X^*$. It now follows from (8.3) and (8.4) that

$$\langle x^*, x \rangle_X \le m_A(x) < 1 \le m_A(x_0) = \langle u^*, x_0 \rangle_U = \langle x^*, x_0 \rangle_X \qquad \text{for all } x \in A,$$

which implies the separation property.

For the general case $0 \notin A$, choose $\tilde{x} \in A$ and consider $\tilde{A} := A - \tilde{x} := \{x - \tilde{x} : x \in A\}$. Then $\tilde{A}$ is open as well and satisfies $0 \in \tilde{A}$ and $x_0 - \tilde{x} \notin \tilde{A}$. As above, we can thus construct an $x^* \in X^*$ with $\langle x^*, y \rangle_X < \langle x^*, x_0 - \tilde{x} \rangle_X$ for all $y \in \tilde{A}$. For all $x \in A$, we then have by definition that $x - \tilde{x} \in \tilde{A}$ and therefore again

$$\langle x^*, x \rangle_X = \langle x^*, x - \tilde{x} \rangle_X + \langle x^*, \tilde{x} \rangle_X < \langle x^*, x_0 - \tilde{x} \rangle_X + \langle x^*, \tilde{x} \rangle_X = \langle x^*, x_0 \rangle_X.$$

Finally, the case $\mathbb{F} = \mathbb{C}$ is reduced to the real case as in the proof of Theorem 8.2.   □

Theorem 8.9 can be used to obtain further separation theorems that are of fundamental importance in convex optimization. We first consider the separation of two disjoint convex sets.

Theorem 8.10
Let X be a normed vector space and let $A_1, A_2 \subset X$ be nonempty and convex with $A_1 \cap A_2 = \emptyset$. If A_1 is open, then there exist an $x^* \in X^*$ and an $\alpha \in \mathbb{R}$ such that

$$\mathrm{Re}\,\langle x^*, x_1 \rangle_X < \alpha \le \mathrm{Re}\,\langle x^*, x_2 \rangle_X \qquad \text{for all } x_1 \in A_1, \ x_2 \in A_2.$$

Proof. Set $A := A_1 - A_2 = \{x_1 - x_2 : x_1 \in A_1, x_2 \in A_2\}$. Then A is open, since $x \in A$ implies that $x \in A_1 - x_2 \subset A$ for some $x_2 \in A_2$ and $A_1 - x_2$ is open. Furthermore, $0 \notin A$ since $A_1 \cap A_2 = \emptyset$. The Hahn–Banach separation theorem (Theorem 8.9) thus yields an $x^* \in X^*$ with $\mathrm{Re}\,\langle x^*, x \rangle_X < \mathrm{Re}\,\langle x^*, 0 \rangle_X = 0$ for all $x \in A$, i.e.,

$$\mathrm{Re}\,\langle x^*, x_1 \rangle_X - \mathrm{Re}\,\langle x^*, x_2 \rangle_X = \mathrm{Re}\,\langle x^*, x_1 - x_2 \rangle_X < 0 \qquad \text{for all } x_1 \in A_1, \ x_2 \in A_2.$$

The claim now follows with $\alpha := \sup_{x_1 \in A_1} \mathrm{Re}\,\langle x^*, x_1 \rangle_X$, since the image of the open set A_1 under the surjective linear operator $x^* \in L(X, \mathbb{R}) \setminus \{0\}$ is open by the open mapping theorem (Theorem 5.5), and hence the supremum is not attained.   □

Another variant *strictly* separates a point from a *closed* set.

Theorem 8.11
Let X be a normed vector space, $A \subset X$ nonempty, closed, and convex, and let $x_0 \in X \setminus A$. Then there exist an $x^* \in X^*$ and an $\alpha \in \mathbb{R}$ such that

$$\mathrm{Re}\,\langle x^*, x \rangle_X \le \alpha < \mathrm{Re}\,\langle x^*, x_0 \rangle_X \qquad \text{for all } x \in A.$$

Proof. Since A is closed, there exists an open (convex) ball $U_\varepsilon(x_0) \subset X \setminus A$. Applying Theorem 8.10 to $A_1 = U_\varepsilon(x_0)$ and $A_2 = A$, we obtain a separating functional $x^* \in X^*$ with $\operatorname{Re} \langle x^*, x_1 \rangle_X < \operatorname{Re} \langle x^*, x_2 \rangle_X$ for all $x_1 \in U_\varepsilon(x_0)$ and $x_2 \in A$. Since we can write every $x_1 \in U_\varepsilon(x_0)$ as $x_1 = x_0 + \varepsilon y$ for some $y \in B_X$, this implies that

$$\operatorname{Re} \langle x^*, x_0 \rangle_X + \varepsilon \operatorname{Re} \langle x^*, y \rangle_X = \operatorname{Re} \langle x^*, x_1 \rangle_X < \operatorname{Re} \langle x^*, x \rangle_X \quad \text{for all } y \in B_X, x \in A.$$

Taking the supremum over all $y \in B_Y$ and using the fact that $x^* \neq 0$ and $\|\operatorname{Re} x^*\|_{(X_\mathbb{R})^*} = \|x^*\|_{X^*}$ (see the proof of Theorem 8.9) yields

$$\operatorname{Re} \langle x^*, x_0 \rangle_X < \operatorname{Re} \langle x^*, x_0 \rangle_X + \varepsilon \|x^*\|_{X^*} \leq \operatorname{Re} \langle x^*, x \rangle_X \quad \text{for all } x \in A.$$

We thus obtain the claim for $-x^*$ and $-\alpha := \operatorname{Re} \langle x^*, x_0 \rangle_X + \varepsilon \|x^*\|_X$. $\qquad\square$

Problems

Problem 8.1 *(Extension of functionals)*
Consider the limit as a linear functional on $c(\mathbb{F})$, i.e.,

$$f(x) = \lim_{k \to \infty} x_k \quad \text{for all } x = \{x_k\}_{k \in \mathbb{N}} \in c(\mathbb{F}).$$

Show that there exist two different continuous extensions $f_1, f_2 \in \ell^\infty(\mathbb{F})^*$ of f with

$$f_1|_{c(\mathbb{F})} = f_2|_{c(\mathbb{F})} = f \quad \text{and} \quad \|f_1\|_{\ell^\infty(\mathbb{F})^*} = \|f_2\|_{\ell^\infty(\mathbb{F})^*} = \|f\|_{c(\mathbb{F})^*}.$$

Hint: Consider the sequence $a = (0, 1, 0, 1, \dots)$, *i.e.,* $a_k = 1$ *for k even and* $a_k = 0$ *for k odd, as well as*

$$U = \{x + \lambda a : x \in c(\mathbb{F}), \lambda \in \mathbb{F}\}.$$

Problem 8.2 *(Functional problem)*
Let X be a normed vector space and I an arbitrary set. Let $x_i \in X$ and $c_i \in \mathbb{F}$ for all $i \in I$ satisfy for some $M > 0$ that

$$\left| \sum_{i \in F} \lambda_i c_i \right| \leq M \left\| \sum_{i \in F} \lambda_i x_i \right\| \quad \text{for all } \lambda_i \in \mathbb{F} \text{ and all finite } F \subset I.$$

Show that there exists a unique $x^* \in X^*$ such that $x^*(x_i) = c_i$ for all $i \in I$. (In this case, $\|x^*\|_{X^*} \leq M$.)

Problem 8.3 *(Infinite systems of linear equations)*
Consider the system of linear equations

$$\sum_{k=1}^{\infty} a_{jk}x_k = b_j, \qquad 1 \le j < \infty,$$

with $|a_{jk}| < \infty$ for all $1 \le j, k < \infty$. Show that this system has a unique solution $x \in \ell^p(\mathbb{F})$ with $\|x\|_p \le M$ if and only if

$$\left| \sum_{k=1}^{n} c_j b_j \right| \le M \left(\sum_{k=1}^{\infty} \left| \sum_{k=1}^{n} c_j a_{jk} \right| \right) \qquad \text{for all } c_j \in \mathbb{F} \text{ and } n \in \mathbb{N}.$$

Problem 8.4 *(Closure and interior of convex sets)*
Let X be a normed vector space and let $A \subset X$ be convex. Show that

(i) cl A and int A are convex;
(ii) if int $A \ne \emptyset$, then cl $A = $ cl int A.

Problem 8.5 *(Convex sets under linear mappings)*
Let X and Y be normed vector spaces and let $T : X \to Y$ be linear. Show that

(i) if $A \subset X$ is convex, then $T(A)$ is convex as well;
(ii) if $B \subset Y$ is convex, then $T^{-1}(B)$ is convex as well.

Problem 8.6 *(Dual space of $\ell^\infty(\mathbb{F})$)*

(i) Show that

$$T : \ell^1(\mathbb{F}) \to \ell^\infty(\mathbb{F})^*, \qquad \langle Tx, y \rangle_{\ell^\infty} = \sum_{k=1}^{\infty} x_n y_n,$$

is not surjective by constructing a functional $x^* \in \ell^\infty(\mathbb{F})^*$ that has no preimage under T.
(ii) Show that every $x^* \in \ell^\infty(\mathbb{F})^*$ can be uniquely represented as the sum of two functionals $x_1^*, x_2^* \in \ell^\infty(\mathbb{F})^*$ with $\langle x_1^*, y \rangle_\infty = \sum_{k=1}^{\infty} x_k y_k$ for some $x \in \ell^1(\mathbb{F})$ and $x_2^*|_{c_0(\mathbb{F})} = 0$. *Hint: Consider $\langle x^*, e_k \rangle_\infty$.*

Problem 8.7 *(Quotient spaces)*
Let X be a normed vector space and $U \subset X$ a closed subspace. Show that

$$T : X^*/U^\perp \to U^*, \qquad x^* + U^\perp \mapsto x^*|_U,$$

is an isometric isomorphism.

Adjoint Operators

9

Just as a normed vector space is characterized by its dual space, a linear operator is characterized by a "dual operator". We define for a continuous linear operator $T \in L(X, Y)$ between normed vector spaces X, Y the *adjoint operator*

$$T^* : Y^* \to X^*, \qquad \langle T^* y^*, x \rangle_X := \langle y^*, T x \rangle_Y \qquad \text{for all } x \in X, \ y^* \in Y^*.$$

By construction, T^* is linear; it is also continuous.

Lemma 9.1

Let $T \in L(X, Y)$. Then $T^* \in L(Y^*, X^*)$ with $\|T^*\|_{L(Y^*, X^*)} = \|T\|_{L(X, Y)}$.

Proof. By definition of the operator norm,

$$\|T^*\|_{L(Y^*, X^*)} = \sup_{y^* \in B_{Y^*}} \|T^* y^*\|_{X^*} = \sup_{y^* \in B_{Y^*}} \sup_{x \in B_X} |\langle T^* y^*, x \rangle_X|$$

$$= \sup_{y^* \in B_{Y^*}} \sup_{x \in B_X} |\langle y^*, T x \rangle_Y| = \sup_{x \in B_X} \sup_{y^* \in B_{Y^*}} |\langle y^*, T x \rangle_Y|$$

$$= \sup_{x \in B_X} \|T x\|_Y = \|T\|_{L(X, Y)},$$

where we have used Corollary 8.4 for the next-to-last equality. $\qquad\square$

© Springer Nature Switzerland AG 2020
C. Clason, *Introduction to Functional Analysis*, Compact Textbooks
in Mathematics, https://doi.org/10.1007/978-3-030-52784-6_9

Example 9.2 A trivial example is the identity $\mathrm{Id}_X \in L(X, X)$: we have

$$\langle x^*, \mathrm{Id}_X\, x \rangle_X = \langle x^*, x \rangle_X = \langle \mathrm{Id}_{X^*}\, x^*, x \rangle_X \qquad \text{for all } x \in X,\ x^* \in X^*,$$

i.e., $(\mathrm{Id}_X)^* = \mathrm{Id}_{X^*}$.

For a less trivial example, we define on $X = Y = \ell^p(\mathbb{F})$ with $1 \le p < \infty$ the *right-shift operator*

$$S_+ : \ell^p(\mathbb{F}) \to \ell^p(\mathbb{F}), \qquad (x_1, x_2, x_3, \dots) \mapsto (0, x_1, x_2, \dots).$$

Since the adjoint operator $(S_+)^* : \ell^p(\mathbb{F})^* \to \ell^p(\mathbb{F})^*$ operates on linear functionals, it is hard to give an explicit expression. However, we can use that $\ell^p(\mathbb{F})^* \cong \ell^q(\mathbb{F})$ with $\frac{1}{p} + \frac{1}{q} = 1$ to characterize its representation $T_p^{-1}(S_+)^* T_p : \ell^q(\mathbb{F}) \to \ell^q(\mathbb{F})$, where $T_p : \ell^q(\mathbb{F}) \to \ell^p(\mathbb{F})^*$ is the isometric isomorphism from Theorem 7.1. Let $y \in \ell^q(\mathbb{F})$ be given. Then we have for all $x \in \ell^p(\mathbb{F})$ that

$$\langle (S_+)^* T_p y, x \rangle_p = \langle T_p y, S_+ x \rangle_p = \sum_{k=2}^{\infty} x_{k-1} y_k = \sum_{k=1}^{\infty} x_k y_{k+1} = \langle T_p z, x \rangle_p$$

with $z = (y_2, y_3, y_4, \dots) \in \ell^q(\mathbb{F})$. This shows that $(S_+)^* T_p y = T_p z \in \ell^p(\mathbb{F})^*$ and hence that $T_p^{-1}(S_+)^* T_p y = z$. The adjoint operator can thus be represented as the *left-shift operator*

$$S_- : \ell^q(\mathbb{F}) \to \ell^q(\mathbb{F}), \qquad (x_1, x_2, x_3, \dots) \mapsto (x_2, x_3, x_4, \dots).$$

Further examples can be obtained using the following calculus, which follows directly from the definition.

Lemma 9.3
Let X, Y, Z be normed vector spaces, $T_1, T_2 \in L(X, Y)$, and $S \in L(Y, Z)$. Then

(i) $(T_1 + T_2)^* = T_1^* + T_2^*$;
(ii) $(\lambda T_1)^* = \lambda T_1^*$ for all $\lambda \in \mathbb{F}$;
(iii) $(S \circ T)^* = T^* \circ S^*$.

A particularly useful property is the following "commutativity".

Theorem 9.4
Let $T \in L(X, Y)$ be continuously invertible. Then $T^* \in L(Y^*, X^*)$ is continuously invertible as well with

$$(T^*)^{-1} = (T^{-1})^* =: T^{-*}.$$

Proof. If T is continuously invertible, then by definition $T^{-1}T = \text{Id}_X$ as well as $TT^{-1} = \text{Id}_Y$, and T^{-1} is continuous. Taking the adjoint of both sides of the first equality and using Lemma 9.3 (iii) yields that $T^*(T^{-1})^* = \text{Id}_{X^*}$; we argue similarly for TT^{-1}. Hence $(T^{-1})^*$ is the inverse of T^*, which is continuous by Lemma 9.1. $\square$

Corollary 9.5

If X and Y are (isometrically) isomorphic, then X^* and Y^* are (isometrically) isomorphic as well.

Proof. If X and Y are isomorphic, then by definition there exists a continuously invertible operator $T : X \to Y$. Then $T^* : Y^* \to X^*$ is also continuously invertible by Theorem 9.4, and hence Y^* and X^* are isomorphic.

If T is an isometry, then Lemma 9.1 implies that $\|T^*\|_{L(Y^*,X^*)} = \|T\|_{L(X,Y)} = 1$ and hence that

$$\|T^*y^*\|_{X^*} \le \|T^*\|_{L(Y^*,X^*)}\|y^*\|_{Y^*} = \|y^*\|_{Y^*} \quad \text{for all } y^* \in Y^*.$$

Conversely, for all $y^* \in Y^*$,

$$\|y^*\|_{Y^*} = \|T^{-*}T^*y^*\|_Y^* \le \|T^{-*}\|_{L(X^*,Y^*)}\|T^*y^*\|_{X^*}$$

$$= \|T^{-1}\|_{L(Y,X)}\|T^*y^*\|_{X^*} = \|T^*y^*\|_{X^*},$$

where we have used that the inverse of an isometric isomorphism is again an isometric isomorphism (which is easily verified using $y = Tx$ if and only if $x = T^{-1}y$). It follows that $\|T^*y^*\|_{X^*} = \|y^*\|_{Y^*}$, and hence T^* is an isometry. $\square$

We now use the adjoint operator to study the solvability of the operator equation $Tx = y$. If X and Y are Banach spaces, this equation has a unique solution that depends continuously on the right-hand side if and only if T is injective and surjective. For finite-dimensional operators (i.e., if $Tx = y$ is a system of linear equations), this can be expressed via the rank of T: the rank–nullity theorem states that T is surjective if and only if T^* (which in this case corresponds to the transposed matrix) is injective (and vice versa). A similar result would of course be useful for infinite-dimensional spaces as well, for which we use the annihilators from Theorem 7.3 in place of the rank–nullity theorem.

Theorem 9.6
Let X and Y be normed vector spaces and $T \in L(X, Y)$. Then

  (i) $(\operatorname{ran} T)^{\perp} = \ker T^*$;
  (ii) $(\ker T^*)_{\perp} = \operatorname{cl}(\operatorname{ran} T)$;
  (iii) $(\operatorname{ran} T^*)_{\perp} = \ker T$;
  (iv) $(\ker T)^{\perp} \supset \operatorname{cl}(\operatorname{ran} T^*)$.

Proof. For (i), let $y^* \in (\operatorname{ran} T)^{\perp}$. Then we have for all $x \in X$ that

$$0 = \langle y^*, Tx \rangle_X = \langle T^* y^*, x \rangle_X.$$

Hence by definition, $T^* y^* = 0 \in X^*$, i.e., $y^* \in \ker T^*$. Conversely, the same argument shows that $y^* \in \ker T^*$ implies that $\langle y^*, Tx \rangle_X = 0$, i.e., $y^* \in (\operatorname{ran} T)^{\perp}$.

Claim (ii) follows from (i) together with Corollary 8.8, since

$$(\ker T^*)_{\perp} = ((\operatorname{ran} T)^{\perp})_{\perp} = \operatorname{cl}(\operatorname{ran} T).$$

Claims (iii) and (iv) follow analogously (keeping in mind the remark after Corollary 8.8). □

This shows that T is surjective if and only if T^* is injective and $\operatorname{ran} T$ is closed. The latter condition can be characterized via the adjoint as well; this is done using another fundamental theorem of functional analysis, the *closed range theorem*. Its proof, which is based on the Hahn–Banach theorems, requires the following two lemmas.

Lemma 9.7
Let X and Y be Banach spaces and $T \in L(X, Y)$. Then the following properties are equivalent:

  (i) ran T is closed;
  (ii) there is a $C > 0$ such that for all $y \in \operatorname{ran} T$ there exists an $x \in X$ with $Tx = y$ and $\|x\|_X \leq C \|y\|_Y$.

Proof. (i) $\Rightarrow$ *(ii):* We use that $\operatorname{ran} T$ is isomorphic to $X / \ker T$ by Theorem 6.4; this implies that the operator $S : X / \ker T \to \operatorname{ran} T$ from Lemma 6.3 with $T = S \circ Q$, where Q denotes the quotient mapping $x \mapsto [x]$, is continuously invertible. Hence for all $x \in X$ there exists an $[x] \in X/\ker T$ with $S[x] = Tx$ and

$$\|[x]\|_{X/\ker T} \le \|S^{-1}\|_{L(\operatorname{ran} T, X/\ker T)} \|Tx\|_Y.$$

As in the proof of Theorem 6.2, we can now find for given $\varepsilon := \|[x]\|_{X/\ker T}$ a $u_\varepsilon \in \ker T$ with

$$\|x - u_\varepsilon\|_X \le \inf_{u \in \ker T} \|x - u\|_X + \varepsilon = 2\|[x]\|_{X/\ker T}.$$

Hence $\tilde{x} := x - u_\varepsilon \in X$ satisfies $T\tilde{x} = Tx = S[x] = y$ as well as the desired estimate for $C := 2\|S^{-1}\|_{L(\operatorname{ran} T, X/\ker T)}$.

  (ii) $\Rightarrow$ (i): Using the assumption and the definition of S from above, we can find for $y \in \operatorname{ran} T$ a unique $[x] \in X/\ker T$ with $S[x] = Tx = y$ and

$$\|[x]\|_{X/\ker T} \le \|x\|_X \le C\|y\|_Y.$$

Hence S is continuously invertible, which by Corollary 1.9 implies that $\operatorname{ran} T$ is closed as the preimage of the Banach spaces $X/\ker T$ under the continuous mapping S^{-1}. □

Note that the inequality in (ii) has to be satisfied for only *one* preimage, not for all. In fact, the proof shows that (ii) is equivalent to

$$\|[x]\|_{X/\ker T} \le C\|Tx\|_Y \qquad \text{for all } x \in X. \tag{9.1}$$

If the inequality holds for *all* preimages, the operator is injective as well.

Corollary 9.8

Let X and Y be Banach spaces and $T \in L(X, Y)$. If there exists $C > 0$ such that

$$\|x\|_X \le C\|Tx\|_Y \qquad \text{for all } x \in X,$$

then T is injective and $\operatorname{ran} T$ is closed.

Proof. If $Tx = 0$, the assumed inequality yields $x = 0$ and hence injectivity of T. In this case, x is the only preimage of Tx; hence Lemma 9.7 (ii) holds and therefore $\operatorname{ran} T$ is closed. □

In this case, all that is missing for the continuous invertibility of T is surjectivity, which can be verified using the following lemma.

Lemma 9.9

Let X and Y be Banach spaces and $T \in L(X, Y)$. If there exists a $c > 0$ such that

$$c\|y^*\|_{Y^*} \leq \|T^*y^*\|_{X^*} \qquad \text{for all } y^* \in Y^*,$$

then T is surjective.

Proof. We use the open mapping theorem (Theorem 5.5). To this end, we show that $\delta U_Y \subset T(U_X)$ for some $\delta > 0$ and the open unit balls U_X, U_Y in X and Y, respectively; as in the proof of Theorem 5.5, it suffices to show that $cU_Y \subset \operatorname{cl} T(U_X) =: A$. We argue by contraposition. Let $y_0 \notin A$ be arbitrary. Since A is nonempty, convex, and closed, the strict separation theorem (Theorem 8.11) yields a $y^* \in Y^*$ and an $\alpha \in \mathbb{R}$ such that

$$\operatorname{Re}\langle y^*, y \rangle_Y \leq \alpha < \operatorname{Re}\langle y^*, y_0 \rangle_Y \qquad \text{for all } y \in A.$$

By linearity of T, if $y \in A$, then also $\sigma y \in A$ for all $\sigma \in \mathbb{F}$ with $|\sigma| = 1$. Hence

$$|\langle y^*, y \rangle_Y| = \operatorname{Re}\langle y^*, \sigma y \rangle_Y < \operatorname{Re}\langle y^*, y_0 \rangle_Y \leq |\langle y^*, y_0 \rangle_Y| \qquad \text{for all } y \in A,$$

where we have chosen $\sigma \in \mathbb{F}$ such that $\sigma \langle y^*, y \rangle_Y = |\langle y^*, y \rangle_Y|$ and $|\sigma| = 1$.

Using the assumption and Lemma 4.3 (i) then shows that

$$c\|y^*\|_{Y^*} \leq \|T^*y^*\|_{X^*} = \sup_{x \in U_X} |\langle T^*y^*, x \rangle_X| = \sup_{x \in U_X} |\langle y^*, Tx \rangle_Y|$$

$$\leq |\langle y^*, y_0 \rangle_Y| \leq \|y^*\|_{Y^*} \|y_0\|_Y$$

and hence that $\|y_0\|_Y \geq c$, i.e., $y_0 \notin cU_Y$. By contraposition, we obtain that $cU_Y \subset A = \operatorname{cl} T(U_X)$. □

We now arrive at the main theorem of this chapter.

Theorem 9.10 (Closed range)

Let X and Y be Banach spaces and $T \in L(X, Y)$. Then the following properties are equivalent:

  (i) $\operatorname{ran} T$ is closed;
  (ii) $\operatorname{ran} T = (\ker T^*)_\perp$;
  (iii) $\operatorname{ran} T^*$ is closed;
  (iv) $\operatorname{ran} T^* = (\ker T)^\perp$.

Proof. (i) ⇔ (ii): If ran T is closed, then Theorem 9.6 (ii) yields

$$(\ker T^*)_\perp = \mathrm{cl}(\mathrm{ran}\, T) = \mathrm{ran}\, T.$$

The other direction is an immediate consequence of the closedness of annihilators.

(i) ⇒ (iv): First, we always have that $\langle T^*y^*, x\rangle_X = \langle y^*, Tx\rangle_Y = 0$ for all $x \in \ker T$ and hence that ran $T^* \subset (\ker T)^\perp$. To show the reverse inclusion, we construct for $x^* \in (\ker T)^\perp$ a $y^* \in Y^*$ with $x^* = T^*y^*$ using the Hahn–Banach extension theorem (Theorem 8.2). We first define the linear functional

$$y_0^* : \mathrm{ran}\, T \to \mathbb{F}, \qquad \langle y_0^*, Tx\rangle_{\mathrm{ran}\, T} := \langle x^*, x\rangle_X \quad \text{for all } x \in X.$$

This functional is well-defined, since $Tx_1 = Tx_2$ for $x_1, x_2 \in X$ implies that $x_1 - x_2 \in \ker T$ and hence that $\langle x^*, x_1 - x_2\rangle_X = 0$. The continuity of y_0^* follows from Lemma 9.7: since ran T is closed, there exists a $C > 0$ such that $\|x\|_X \leq C\|y\|_Y$ for all $y \in \mathrm{ran}\, T$ and some $x \in X$ with $Tx = y$. For every $y \in \mathrm{ran}\, T$ and accordingly chosen $x \in X$, we thus have

$$|\langle y_0^*, y\rangle_{\mathrm{ran}\, T}| = |\langle y_0^*, Tx\rangle_{\mathrm{ran}\, T}| = |\langle x^*, x\rangle_X| \leq \|x^*\|_{X^*}\|x\|_X \leq C\|x^*\|_{X^*}\|y\|_Y.$$

Hence y_0^* is continuous. Since ran $T \subset Y$ is a subspace, we can use Theorem 8.2 to extend y_0^* to a continuous linear functional $y^* \in Y^*$. It follows that

$$\langle x^*, x\rangle_X = \langle y_0^*, Tx\rangle_{\mathrm{ran}\, T} = \langle y^*, Tx\rangle_Y = \langle T^*y^*, x\rangle_X \qquad \text{for all } x \in X$$

and hence that $x^* = T^*y^*$, i.e., $x^* \in \mathrm{ran}\, T^*$.

(iv) ⇒ (iii) again follows directly from the closedness of annihilators.

(iii) ⇒ (i): We show that ran $T = \mathrm{cl}(\mathrm{ran}\, T) =: U$. We do this by constructing an operator $S \in L(X, U)$ via $Sx := Tx$ for all $x \in X$ such that ran $S = \mathrm{ran}\, T \subset U$ is dense and then showing that S is surjective (i.e., ran $T = \mathrm{ran}\, S = U$). First, denoting the restriction of $y^* \in Y^*$ to $U \subset Y$ by $y^*|_U \in U^*$, we have

$$\langle T^*y^*, x\rangle_X = \langle y^*, Tx\rangle_Y = \langle y^*|_U, Sx\rangle_U = \langle S^*y^*|_U, x\rangle_X \qquad \text{for all } x \in X,$$

i.e., $T^*y^* = S^*y^*|_U$ and hence ran $T^* \subset \mathrm{ran}\, S^*$. Conversely, let $u^* \in U^*$ and hence $S^*u^* \in \mathrm{ran}\, S^*$ be arbitrary. We extend u^* using Theorem 8.2 to $y^* \in Y^*$; as above, we then have

$$\langle S^*u^*, x\rangle_X = \langle u^*, Sx\rangle_U = \langle y^*, Tx\rangle_Y = \langle T^*y^*, x\rangle_X \qquad \text{for all } x \in X,$$

i.e., $S^*u^* = T^*y^*$ and hence ran $S^* \subset \mathrm{ran}\, T^*$. Since ran S is by construction dense in U, it follows from Corollary 8.6, Theorem 9.6 (ii), and Corollary 8.8 that

$$\{0\} = (\mathrm{ran}\, S)^\perp = ((\ker S^*)_\perp)^\perp = \mathrm{cl}\,\ker S^* = \ker S^*.$$

Hence S^* is injective. Furthermore, ran $S^* = \text{ran } T^*$ is closed by assumption and therefore S^* is continuously invertible by Corollary 5.7. We thus have for all $u^* \in U^*$ that

$$\|u^*\|_{U^*} = \|S^{-*}S^*u^*\|_{U^*} \leq \|S^{-*}\|_{L(X^*, U^*)} \|S^*u^*\|_{X^*}.$$

This verifies the assumption of Lemma 9.9 with $c := \|S^{-*}\|_{L(X^*, U^*)}^{-1}$. Hence S is surjective, showing that ran $S = U = \text{cl(ran } T)$, i.e., ran $T = \text{ran } S$ is closed. □

In practice, the closed range theorem is often applied by verifying the assumption of Corollary 9.8 as well as the injectivity of T^*; the former then yields injectivity of T and the latter surjectivity together with the closedness of the range. It follows that the operator equation $Tx = y$ has a unique solution for all $y \in Y$, which moreover satisfies the *a priori estimate* $\|x\|_X \leq C\|y\|_Y$. This approach is a fundamental tool in the theory of partial differential equations.

Problems

Problem 9.1 *(Examples of adjoint operators)*
Determine the adjoints of

(i) $A : \ell^1(\mathbb{R}) \to \ell^1(\mathbb{R})$, $\quad \{x_k\}_{k \in \mathbb{N}} \mapsto \left(\sum\limits_{k=1}^{\infty} x_k, 0, 0, \ldots \right)$;

(ii) $B : \ell^2(\mathbb{R}) \to \ell^2(\mathbb{R})$, $\quad \{x_k\}_{k \in \mathbb{N}} \mapsto \left\{ \dfrac{1}{k^2} \sum\limits_{j=1}^{k} x_j \right\}_{k \in \mathbb{N}}$.

Problem 9.2 *(Calculus for adjoint operators (Lemma 9.3))*
Let X, Y, Z be normed vector spaces, $T_1, T_2 \in L(X, Y)$, and $S \in L(Y, Z)$. Show that

(i) $(T_1 + T_2)^* = T_1^* + T_2^*$;
(ii) $(\lambda T_1)^* = \lambda T_1^*$ for all $\lambda \in \mathbb{F}$;
(iii) $(S \circ T_1)^* = T_1^* \circ S^*$.

Problem 9.3 *(Continuous embeddings)*
Let X and Y be normed vector spaces such that $X \hookrightarrow Y$ is dense in Y. Show that the *restriction operator*

$$R : Y^* \to X^*, \qquad y^* \mapsto y^*|_X,$$

is continuous and injective (i.e., that $Y^* \hookrightarrow X^*$).

Problem 9.4 *(Adjoint operators)*

Let X and Y be Banach spaces. Show that both A and B are linear and continuous for

(i) $A : X \to Y$ and $B : Y^* \to X^*$ with

$$\langle y^*, A(x) \rangle_Y = \langle B(y^*), x \rangle_X \quad \text{for all } x \in X \text{ and } y^* \in Y^*;$$

(ii) $A : X \to Y^*$ and $B : Y \to X^*$ with

$$\langle A(x), y \rangle_Y = \langle B(y), x \rangle_X \quad \text{for all } x \in X \text{ and } y \in Y.$$

Problem 9.5 *(Banach–Nečas–Babuška theorem)*

Let X and Y be Banach spaces. Show that $T \in L(X, Y)$ is continuously invertible if and only if it satisfies the following properties:

(i) there exists an $\alpha > 0$ such that

$$\|Tx\|_Y \geq \alpha \|x\|_X \quad \text{for all } x \in X;$$

(ii) $\ker T^* = \{0\}$.

Show that these assumptions further imply that $\|T^{-1}\|_{L(Y,X)} \leq \alpha^{-1}$.

Reflexivity

10

We have seen in the previous chapters that the dual space X^* of a normed vector space X is useful since it characterizes X in a suitable way. Similarly, X^* itself is characterized by *its* dual space $(X^*)^*$; a natural question is now whether this is transitive, i.e., whether the *bidual space* $X^{**} := (X^*)^*$ characterizes X directly.

The first step is to study the elements of X^{**}. Consider for $x \in X$ and $x^* \in X^*$ the duality pairing $\langle x^*, x \rangle_X \in \mathbb{F}$. For fixed x^*, this defines a continuous linear mapping from X to $\mathbb{F}$ (which coincides with x^* by the definition of the duality pairing). Conversely, fixing x defines a linear mapping x^{**} from X^* to $\mathbb{F}$, which is also continuous since

$$x^{**}(x^*) := \langle x^*, x \rangle_X \leq \|x^*\|_{X^*} \|x\|_X \qquad \text{for all } x^* \in X^*. \tag{10.1}$$

In this way, we have constructed a *canonical embedding*

$$J_X : X \to X^{**}, \qquad \langle J_X(x), x^* \rangle_{X^*} = \langle x^*, x \rangle_X \quad \text{for all } x^* \in X^*.$$

It is straightforward to verify that J_X is linear and continuous by (10.1). Furthermore, it follows from Corollary 8.4 that

$$\|x\|_X = \max_{x^* \in B_{X^*}} |\langle x^*, x \rangle_X| = \max_{x^* \in B_{X^*}} |\langle J_X x, x^* \rangle_{X^*}| = \|J_X x\|_{X^{**}},$$

and hence J_X is an isometry and therefore injective. We have thus shown

---

Theorem 10.1

The canonical embedding $J_X : X \to X^{**}$ is linear, injective, and isometric.

© Springer Nature Switzerland AG 2020
C. Clason, *Introduction to Functional Analysis*, Compact Textbooks in Mathematics, https://doi.org/10.1007/978-3-030-52784-6_10

By linearity, ran J_X is a subspace of X^{**}. Since J_X is an isometry and X^{**} is a Banach space (by virtue of being a dual space), ran J_X is complete if and only if X is complete. By Lemma 3.5, the former is the case if and only if ran J_X closed in X^{**}. If X is not complete, we can take cl ran $J_X \subset X^{**}$ as a "completion" of X.

In general, ran J_X is a proper subspace since J_X need not be surjective. If J_X is surjective, then X is called *reflexive*. In this case, J_X is even an isometric isomorphism, which shows that $X \cong X^{**}$. Note that $X \cong X^{**}$ alone does *not* imply that X is reflexive; reflexivity requires that the isometric isomorphism be specifically given by the canonical embedding.

Example 10.2

(i) All finite-dimensional vector spaces are reflexive since

$$\dim(X^{**}) = \dim(X^*) = \dim(X),$$

and hence the injectivity of J_X already implies the surjectivity.

(ii) The sequence spaces $\ell^p(\mathbb{F})$ are reflexive for $1 < p < \infty$. To show this, we use the isometries $T_p : \ell^q(\mathbb{F}) \to \ell^p(\mathbb{F})^*$ and $T_q : \ell^p(\mathbb{F}) \to \ell^q(\mathbb{F})^*$ from Theorem 7.1, which satisfy

$$\langle T_q x, y \rangle_{\ell^q} = \sum_{k=1}^{\infty} x_k y_k = \langle T_p y, x \rangle_{\ell^p} \qquad \text{for all } x \in \ell^p(\mathbb{F}), \ y \in \ell^q(\mathbb{F}).$$

Let now $x^{**} \in \ell^p(\mathbb{F})^{**}$ be arbitrary. Then $J_{\ell^p} x = x^{**}$ for $x := T_q^{-1} T_p^* x^{**}$, since we have for arbitrary $x^* \in \ell^p(\mathbb{F})^*$ and $y := T_p^{-1} x^* \in \ell^q(\mathbb{F})$ that

$$\langle J_{\ell^p} x, x^* \rangle_{(\ell^p)^*} = \langle x^*, x \rangle_{\ell^p} = \langle T_p y, x \rangle_{\ell^p} = \langle T_q x, y \rangle_{\ell^q}$$

$$= \langle T_p^* x^{**}, y \rangle_{\ell^q} = \langle x^{**}, T_p y \rangle_{(\ell^p)^*} = \langle x^{**}, x \rangle_{(\ell^p)^*}.$$

Hence $J_{\ell^p} = T_p^{-*} T_q$ is surjective.

(iii) The same argument shows that $L^p(\Omega)$ is reflexive for $1 < p < \infty$.

(iv) On the other hand, $\ell^1(\mathbb{F})$, $\ell^\infty(\mathbb{F})$, $c_0(\mathbb{F})$, and $c(\mathbb{F})$ as well as $L^1(\Omega)$, $L^\infty(\Omega)$, and $C_b(X)$ are not reflexive; this follows from the next three theorems.

In analogy to the bidual space, we can define for $T \in L(X, Y)$ a *biadjoint operator* $T^{**} \in L(X^{**}, Y^{**})$. This definition is compatible with the canonical embedding.

Lemma 10.3
Let X and Y be normed vector spaces and $T \in L(X, Y)$. Then

$$T^{**} \circ J_X = J_Y \circ T.$$

Proof. For all $x \in X$ and $y^* \in Y^*$,

$$\langle T^{**} J_X x, y^* \rangle_{Y^*} = \langle J_X x, T^* y^* \rangle_{X^*} = \langle T^* y^*, x \rangle_X = \langle y^*, Tx \rangle_Y = \langle J_Y Tx, y^* \rangle_{Y^*}. \qquad \square$$

We next show some results on "inheritance" of reflexivity.

Theorem 10.4
Let X be a normed vector space and Y a reflexive Banach space with $X \simeq Y$. Then X is reflexive as well.

Proof. If $T : X \to Y$ is an isomorphism, i.e., continuously invertible, then so are T^* and T^{**} by Theorem 9.4. Lemma 10.3 then implies that J_X is continuously invertible and hence surjective if and only if J_Y is. $\qquad \square$

Theorem 10.5
Let X be a reflexive Banach space and $U \subset X$ a closed subspace. Then U is reflexive as well.

Proof. Let $u^{**} \in U^{**}$ be arbitrary. Since every continuous linear functional $x^* \in X^*$ defines a functional $x^*|_U \in U^*$ by restriction, we can define an $x^{**} \in X^{**}$ via

$$\langle x^{**}, x^* \rangle_{X^*} = \langle u^{**}, x^*|_U \rangle_{U^*} \qquad \text{for all } x^* \in X^*. \tag{10.2}$$

Since X is reflexive, there exists an $x := J_X^{-1} x^{**} \in X$. We now show that $x \in U$. Assume to the contrary that $x \notin U$. Then Corollary 8.5 yields an $x^* \in U^\perp$ such that $\langle x^*, x \rangle_X \neq 0$. In particular, $x^*|_U = 0$, which together with (10.2) yields the contradiction

$$0 = \langle u^{**}, x^*|_U \rangle_{U^*} = \langle x^{**}, x^* \rangle_{X^*} = \langle x^*, x \rangle_X \neq 0.$$

Hence $x \in U$, and it remains to show that $J_U x = u^{**}$. To this end, let $u^* \in U^*$ be given. The Hahn–Banach extension theorem (Theorem 8.2) then yields an $x^* \in X^*$ such that $x^*|_U = u^*$.

We thus obtain that

$$\langle J_U x, u^* \rangle_{U^*} = \langle u^*, x \rangle_X = \langle x^*, x \rangle_X = \langle J_X x, x^* \rangle_{X^*}$$
$$= \langle x^{**}, x^* \rangle_{X^*} = \langle u^{**}, x^* |_U \rangle_{U^*} = \langle u^{**}, u^* \rangle_{U^*},$$

i.e., $J_U x = u^{**}$. □

Theorem 10.6

Let X be a Banach space. Then X is reflexive if and only if X^* is reflexive.

Proof. Let X be reflexive. We have to show that $J_{X^*} : X^* \to X^{***}$ is surjective. To this end, let $x^{***} \in X^{***}$ be given and set $x^* := J_X^* x^{***} \in X^*$. Since X is reflexive, there exists for every $x^{**} \in X^{**}$ an $x := J_X^{-1} x^{**} \in X$. This implies that

$$\langle x^{***}, x^{**} \rangle_{X^{**}} = \langle x^{***}, J_X x \rangle_{X^{**}} = \langle J_X^* x^{***}, x \rangle_X = \langle x^*, x \rangle_X$$
$$= \langle J_X x, x^* \rangle_{X^*} = \langle x^{**}, x^* \rangle_{X^*} = \langle J_{X^*} x^*, x^{**} \rangle_{X^{**}},$$

i.e., $x^{***} = J_{X^*} x^*$.

Conversely, let X^* be reflexive. The first part of the proof then shows that X^{**} is reflexive as well. Hence the closed (since X is complete) subspace ran $J_X \subset X^{**}$ is also reflexive by Theorem 10.5. Since X and ran J_X are isometrically isomorphic (since $J_X : X \to \operatorname{ran} J_X$ is surjective by construction and hence an isomorphism by Theorem 10.1), X is reflexive by Theorem 10.4. □

As already claimed, it follows that $\ell^1(\mathbb{F})$ cannot be reflexive: otherwise, $\ell^\infty(\mathbb{F})^* \cong (\ell^1(\mathbb{F})^*)^* \cong \ell^1(\mathbb{F})$ would be separable by Theorem 3.15, implying that also $\ell^\infty(\mathbb{F})$ is separable by Corollary 8.7. But this is not the case by, again, Theorem 3.15. It follows from Theorem 10.6 that $\ell^\infty(\mathbb{F})$ and $c_0(\mathbb{F})$ are also not reflexive since $\ell^\infty(\mathbb{F}) \cong \ell^1(\mathbb{F})^*$ and $c_0(\mathbb{F})^* \cong l^1(\mathbb{F})$. Finally, $c_0(\mathbb{F})$ is a nonreflexive closed subspace of $c(\mathbb{F})$, which thus cannot be reflexive either by Theorem 10.5. The claims for the function spaces follow analogously.

Problems

Problem 10.1 *(Operator norm in reflexive spaces)*
Let X be a reflexive Banach space. Show that

$$\|x^*\|_{X^*} = \max_{x \in B_X} |\langle x^*, x \rangle_X| \qquad \text{for all } x^* \in X^*,$$

i.e., that the supremum in the definition of the operator norm is always attained.

Problem 10.2 *(Operator norm in nonreflexive spaces)*
Let $x^* \in c_0(\mathbb{R})$ be defined via

$$\langle x^*, x \rangle_X := \sum_{k=1}^{\infty} 2^{-k} x_k \qquad \text{for all } x \in c_0(\mathbb{R}).$$

Show that for this x^*, the supremum in the definition of the operator norm is *not* attained.

Problem 10.3 *(Reflexivity of quotient spaces)*
Let X be a reflexive Banach space and $U \subset X$ a closed subspace. Show that X/U is reflexive as well.

Problem 10.4 *(Nonadjoint operators)*

(i) Let X and Y be normed vector spaces. Show that $S \in L(Y^*, X^*)$ is an adjoint operator if and only if

$$\operatorname{ran}(S^* \circ J_X) \subset \operatorname{ran} J_Y.$$

(ii) Give an example for a continuous linear operator that is not an adjoint operator.

Weak Convergence

We finally arrive at the promised generalization of componentwise convergence to infinite-dimensional spaces that allows obtaining a result similar to the Heine–Borel theorem (Theorem 2.5). As argued in Chap. 7, continuous linear functionals are the infinite-dimensional counterparts to the component mappings, and we define our new notion of convergence accordingly.

Let X be a normed vector space and $\{x_n\}_{n\in\mathbb{N}} \subset X$ a sequence. We say that $\{x_n\}_{n\in\mathbb{N}}$ *converges weakly* in X to $x \in X$ and write $x_n \rightharpoonup x$ if

$$\lim_{n\to\infty} \langle x^*, x_n\rangle_X = \langle x^*, x\rangle_X \qquad \text{for all } x^* \in X^*.$$

Similarly, we say that the sequence $\{x_n^*\}_{n\in\mathbb{N}} \subset X^*$ *converges weakly-** in X^* and write $x_n^* \rightharpoonup^* x^*$ if

$$\lim_{n\to\infty} \langle x_n^*, x\rangle_X = \langle x^*, x\rangle_X \qquad \text{for all } x \in X.$$

These limits are unique. For the weak-$*$ convergence, this follows directly from the definition: if $x_n^* \rightharpoonup^* x^*$ and $x_n^* \rightharpoonup^* y^*$, then

$$\langle x^*, x\rangle_X = \lim_{n\to\infty} \langle x_n^*, x\rangle_X = \langle y^*, x\rangle_X \qquad \text{for all } x \in X$$

and hence $x^* = y^*$ by definition. Similarly, $x_n \rightharpoonup x$ and $x_n \rightharpoonup y$ implies that

$$\langle x^*, x\rangle_X = \langle x^*, y\rangle_X \qquad \text{for all } x^* \in X^*.$$

If now $x \neq y$, we obtain from Corollary 8.5 an $x^* \in X^*$ with $\langle x^*, x - y\rangle_X \neq 0$, in contradiction to the above equality. If X is reflexive, then weak convergence

© Springer Nature Switzerland AG 2020
C. Clason, *Introduction to Functional Analysis*, Compact Textbooks
in Mathematics, https://doi.org/10.1007/978-3-030-52784-6_11

(in X) and weak-$*$ convergence (in X^{**}) coincide; in finite-dimensional spaces, all convergences are equivalent.

For a better distinction, we say that a sequence *converges strongly* if it converges with respect to the metric induced by the norm. Every strongly convergent sequence converges weakly; this follows directly from

$$\langle x^*, x_n - x \rangle_X \leq \|x^*\|_{X^*} \|x_n - x\|_X \to 0.$$

Similarly, strong convergence in X^* implies weak-$*$ convergence. However, the converse is not true, as the following example shows.

Example 11.1 Consider the sequence $\{e_n\}_{n\in\mathbb{N}} \subset \ell^p(\mathbb{F})$, $1 \leq p \leq \infty$, of unit vectors. Then $\|e_n\|_p = 1$ for all $n \in \mathbb{N}$, and $\{e_n\}_{n\in\mathbb{N}}$ is not a Cauchy sequence for any p.

For $1 < p < \infty$, we can now use the representation $\ell^p(\mathbb{F})^* \cong \ell^q(\mathbb{F})$: for every $x^* \in \ell^p(\mathbb{F})^*$, there exists a $y \in \ell^q(\mathbb{F})$ such that

$$\langle x^*, e_n \rangle_{\ell^p} = \sum_{k=1}^{\infty} y_k (e_n)_k = y_n \to 0,$$

since $\|y\|_q < \infty$ implies that $\{y_k\}_{k\in\mathbb{N}}$ is a null sequence. Hence $e_n \rightharpoonup 0$ in $\ell^p(\mathbb{F})$ for $1 < p < \infty$, and therefore also $e_n \rightharpoonup^* 0$ by reflexivity.

Since $\ell^1(\mathbb{F})$ is not reflexive, we have a choice here: if we consider $e_n \in \ell^1(\mathbb{F})$ via the isomorphism T from Theorem 7.1 as an element of $c_0(\mathbb{F})^*$, we have

$$\langle Te_n, y \rangle_{c_0} = \sum_{k=1}^{\infty} (e_n)_k y_k = y_n \to 0 \qquad \text{for all } y \in c_0(\mathbb{F})$$

and hence $e_n \rightharpoonup^* 0$ in $\ell^1(\mathbb{F})$. On the other hand, we do *not* have $e_n \rightharpoonup 0$, since the constant sequence $y := \{1\}_{k\in\mathbb{N}} \in \ell^\infty(\mathbb{F}) \cong \ell^1(\mathbb{F})^*$ satisfies

$$\langle Ty, e_n \rangle_{\ell^1} = \sum_{k=1}^{\infty} (e_n)_k y_k = y_n = 1 \qquad \text{for all } n \in \mathbb{N}.$$

But $c_0(\mathbb{F}) \subset \ell^\infty(\mathbb{F})$, and hence the only candidate for the weak limit is 0. This shows that $\{e_n\}_{n\in\mathbb{N}}$ cannot converge weakly at all.[1]

Hence weakly convergent sequences do not necessarily converge strongly; we don't even have $\|x_n\|_X \to \|x\|_X$. However, the following weaker properties hold.

---

[1] In fact, weak convergence is equivalent to strong convergence in $\ell^1(\mathbb{F})$. This is a special property of this space known as *Schur's lemma*; see, e.g., [7, Proposition 5.2].

Theorem 11.2

Let X be a normed vector space, $\{x_n\}_{n\in\mathbb{N}} \subset X$, and $\{x_n^*\}_{n\in\mathbb{N}} \subset X^*$. Then

(i) $x_n \rightharpoonup x$ implies that $\|x\|_X \leq \liminf_{n\to\infty} \|x_n\|_X$;
(ii) $x_n^* \rightharpoonup^* x^*$ implies that $\|x^*\|_{X^*} \leq \liminf_{n\to\infty} \|x_n^*\|_{X^*}$.

Proof. We first show (ii). Let $\{x_n^*\}_{n\in\mathbb{N}} \subset X^*$ be a weakly-$*$ convergent sequence with $x_n^* \rightharpoonup^*$ $x^* \in X^*$. It follows using the reverse triangle inequality that $|\langle x_n^*, x\rangle_X| \to |\langle x^*, x\rangle_X|$ for all $x \in X$ and hence that

$$|\langle x^*, x\rangle_X| = \lim_{n\to\infty} |\langle x_n^*, x\rangle_X| \leq \liminf_{n\to\infty} \|x_n^*\|_{X^*} \|x\|_X.$$

Taking the supremum over all $x \in B_X$ and using the definition of the operator norm then yields the claim.

The same argument using Corollary 8.4 yields (i). □

Property (i) is known as *weak lower semicontinuity*; property (ii) is correspondingly called *weak-$*$ lower semicontinuity*.

Furthermore, weakly convergent sequences are bounded.

Theorem 11.3

Let X be normed vector space. Then every weakly convergent sequence in X is bounded. If X is complete, then every weakly-$*$ convergent sequence in X^* is bounded as well.

Proof. We again first show the claim for X^*. As before, $x_n^* \rightharpoonup^* x^*$ implies that $|\langle x_n^*, x\rangle_X| \to |\langle x^*, x\rangle_X|$ for all $x \in X$ and hence that

$$\sup_{n\in\mathbb{N}} |\langle x_n^*, x\rangle_X| < \infty \qquad \text{for all } x \in X.$$

Since X (by assumption) and X^* (as a dual space) are Banach spaces and the mapping $x^* \mapsto \langle x^*, x\rangle$ is linear and continuous for all $x \in X$, the Banach–Steinhaus theorem (Theorem 5.3) yields

$$\sup_{n\in\mathbb{N}} \|x_n^*\|_{X^*} < \infty,$$

i.e., $\{x_n^*\}_{n\in\mathbb{N}}$ is bounded.

The claim for the weak convergence is reduced to the weak-$*$ convergence (in X^{**}) using the canonical embedding $J_X : X \to X^{**}$. If $\{x_n\}_{n\in\mathbb{N}} \subset X$ converges weakly to some $x \in X$, then

$$\langle J_X x_n - J_X x, x^*\rangle_{X^*} = \langle J_X(x_n - x), x^*\rangle_{X^*} = \langle x^*, x_n - x\rangle_X \to 0 \quad \text{for all } x^* \in X^*,$$

i.e., $\{J_X x_n\}_{n\in\mathbb{N}}$ converges weakly-$*$ in X^{**}. Since J_X is an isometry by Theorem 10.1, the already shown claim yields

$$\sup_{n\in\mathbb{N}} \|x_n\|_X = \sup_{n\in\mathbb{N}} \|J_X x_n\|_{X^{**}} < \infty,$$

i.e., $\{x_n\}_{n\in\mathbb{N}}$ is bounded. $\square$

For example, Theorem 11.3 implies that the duality pairing of weakly(-$*$) convergent sequences converges, as long as one of them converges even strongly.

Corollary 11.4

Let X be a normed vector space, $\{x_n\}_{n\in\mathbb{N}} \subset X$, and $\{x_n^*\}_{n\in\mathbb{N}} \subset X^*$. If either

(i) $x_n \rightharpoonup x$ and $x_n^* \to x^*$ or
(ii) $x_n \to x$ and $x_n^* \rightharpoonup^* x^*$ and X is complete,

then $\langle x_n^*, x_n\rangle_X \to \langle x^*, x\rangle$.

Proof. If (i) holds, we can add and subtract $\langle x^*, x_n\rangle_X$ in order to estimate

$$|\langle x^*, x\rangle_X - \langle x_n^*, x_n\rangle_X| \le |\langle x^*, x - x_n\rangle_X| + |\langle x_n^* - x^*, x_n\rangle_X|$$

$$\le |\langle x^*, x - x_n\rangle_X| + \|x_n\|_X \|x_n^* - x^*\|_{X^*}.$$

On passing to the limit $n \to \infty$, the first summand vanishes since $x_n \rightharpoonup x$; the second summand vanishes since $\{\|x_n\|_X\}_{n\in\mathbb{N}}$ is bounded by Theorem 11.3 and $x_n^* \to x^*$. One argues analogously if (ii) holds. $\square$

However, in general $x_n \rightharpoonup x$ and $x_n^* \rightharpoonup x^*$ do *not* imply that $\langle x_n^*, x_n\rangle_X \to \langle x^*, x\rangle_X$: consider again unit vectors in $\ell^2(\mathbb{F}) \cong \ell^2(\mathbb{F})^*$, which by Example 11.1 satisfy $e_n \rightharpoonup 0$ and $e_n \rightharpoonup^* 0$, but $\langle T e_n, e_n\rangle_{\ell^2} = \sum_{k=1}^{\infty} (e_n)_k^2 = 1$ for all $n \in \mathbb{N}$.

This example also shows that the closed set $S_{\ell^2} := \{x \in \ell^2 : \|x\|_2 = 1\}$ is not *weakly (sequentially) closed*, i.e., $\{x_n\}_{n\in\mathbb{N}} \subset S_{\ell^2}$ with $x_n \rightharpoonup x$ does not imply that $x \in S_{\ell^2}$ (see Theorem 11.2). The reason is that this set is not convex.

Theorem 11.5

Let X be a normed vector space and let $U \subset X$ be convex. Then U is weakly closed if and only U is closed.

Proof. Since a strongly convergent sequence is also weakly convergent, every weakly closed set is also closed.[2] Conversely, let $U \subset X$ be convex and closed, and consider a sequence $\{x_n\}_{n \in \mathbb{N}} \subset U$ with $x_n \rightharpoonup x \in X$. Assume now to the contrary that $x \in X \setminus U$. Then we can use the strict separation theorem (Theorem 8.11) to find an $x^* \in X^*$ and $\alpha \in \mathbb{R}$ such that

$$\mathrm{Re}\langle x^*, x_n \rangle_X \le \alpha < \mathrm{Re}\langle x^*, x \rangle_X \quad \text{for all } n \in \mathbb{N}.$$

Since $x_n \rightharpoonup x$, we can pass to the limit $n \to \infty$ to obtain the contradiction

$$\mathrm{Re}\langle x^*, x \rangle_X \le \alpha < \mathrm{Re}\langle x^*, x \rangle_X. \qquad \square$$

On the other hand, closed convex subsets of X^* are in general *not* weakly-$*$ closed. (An exception is the unit ball B_{X^*} by Theorem 11.2 (ii).)

We now come to the promised results on weak compactness. We call a subset $U \subset X$ *weak sequentially compact* if every sequence in U contains a weakly convergent subsequence with limit in U. We define *weak-$*$ sequential compact* subsets of X^* analogously.

Theorem 11.6 (Banach–Alaoglu[3])

Let X be a normed vector space. If X is separable, then the unit ball B_{X^*} in X^* is weak-$*$ sequentially compact.

Proof. The claim is reminiscent of the Arzelà–Ascoli theorem: continuous linear functionals are elements in $C(X)$, and the boundedness in the operator norm corresponds to equicontinuity. The proof therefore proceeds along the same lines, but in place of the compactness of the metric space we can directly use the separability of X.

---

[2] Weak closedness is therefore a *stronger* property than closedness.

[3] Strictly speaking, this terminology is incorrect: Alaoglu proved the *open cover compactness* of the unit ball in the weak-$*$ topology (and correspondingly his proof is much more involved), while the result shown here was already proved by Banach. For metric spaces, however, the weak-$*$ topology on the unit ball(!) is metrizable, and hence the two results are equivalent. In the literature, one can thus often find this special case under the more general name.

Let $\{x_n^*\}_{n\in\mathbb{N}} \subset B_{X^*}$ be a sequence and $\{x_m : m \in \mathbb{N}\} \subset X$ a dense subset. Then $|\langle x_n^*, x_m\rangle_X| \le \|x_m\|_X$ for all $m \in \mathbb{N}$, and hence the sequences $\{\langle x_n^*, x_m\rangle_X\}_{n\in\mathbb{N}} \subset \mathbb{F}$ are bounded and hence, by completeness of $\mathbb{F}$, each contains a convergent subsequence, whose limits we denote by $\langle x^*, x_m\rangle_X$ (without presupposing the existence of some limit $x^* \in X^*$!) As in the proof of Theorem 2.11, we can then construct a diagonal sequence $\{z_n^*\}_{n\in\mathbb{N}} \subset B_{X^*}$ with $\langle z_n^*, x_m\rangle_X \to \langle x^*, x_m\rangle_X$ for all $m \in \mathbb{N}$. We are now looking for the weak-$*$ limit $x^* \in X^*$ of this diagonal sequence.

To this end, we first define the subspace $Z := \lin\{x_m : m \in \mathbb{N}\}$ as well as the linear functional

$$z^* : Z \to \mathbb{F}, \qquad \langle z^*, z\rangle_X := \lim_{n\to\infty} \langle z_n^*, z\rangle_X \quad \text{for all } z \in Z.$$

This functional is well-defined since all $z \in Z$ are linear combinations of the x_m, and the right-hand side limit can thus be expressed through the corresponding linear combinations of $\langle x^*, x_m\rangle_X$. Furthermore, $|\langle z_n^*, z\rangle_X| \le \|z\|_X$ and hence $|\langle z^*, z\rangle_X| \le \|z\|_X$ for all $z \in Z$, which shows that z^* is continuous with $\|z^*\|_{Z^*} \le 1$. We now extend z^* using the Hahn–Banach extension theorem (Theorem 8.2) to a functional $x^* \in B_{X^*}$.

It remains to show that $x_n^* \rightharpoonup^* x^*$. Let $x \in X$ and $\varepsilon > 0$ be arbitrary. Since Z is dense in X, we can find a $z \in Z$ with $\|z - x\|_X \le \varepsilon$. Furthermore, the convergence $\langle z_n^*, z\rangle_X \to \langle z^*, z\rangle_X$ yields an $N \in \mathbb{N}$ with $|\langle z_n^* - z^*, z\rangle_X| \le \varepsilon$ for all $n \ge N$. We thus obtain for all $n \ge N$ that

$$
\begin{aligned}
|\langle x^*, x\rangle_X - \langle z_n^*, x\rangle_X| &\le |\langle x^* - z_n^*, x - z\rangle_X| + |\langle x^* - z_n^*, z\rangle_X| \\
&\le (\|x^*\|_{X^*} + \|z_n^*\|_{X^*})\|x - z\|_X + |\langle z^* - z_n^*, z\rangle_X| \\
&\le 2\varepsilon + \varepsilon = 3\varepsilon.
\end{aligned}
$$

It follows that $\langle z_n^*, x\rangle_X \to \langle x^*, x\rangle_X$ for all $x \in X$ and hence that $z_n^* \rightharpoonup^* x^*$.   □

Here the separability of X is essential: Consider for example $e_n^* \in \ell^\infty(\mathbb{F})^*$ with $\langle e_n^*, x\rangle_{\ell^\infty} = x_n$. Then $\|e_n^*\|_{(\ell^\infty)^*} = 1$, but since sequences in $\ell^\infty(\mathbb{F})$ need not converge, $\{e_n^*\}_{n\in\mathbb{N}}$ does not contain a weakly-$*$ convergent subsequence.

By a simple scaling argument, it follows from Theorem 11.6 that *every* closed ball in X^* is weak-$*$ sequentially compact. In particular, we obtain a "weak-$*$ Bolzano–Weierstraß theorem".

Corollary 11.7

If X is a separable normed vector space, then every bounded sequence in X^* contains a weakly-$*$ convergent subsequence.

Since every reflexive spaces is isomorphic to its bidual space, we can deduce the weak sequential compactness of the corresponding unit ball.

> **Theorem 11.8 (Eberlein–Šmulian[4])**
>
> Let X be a normed vector space. If X is reflexive, then the unit ball B_X is weak sequentially compact.

Proof. We obtain the claim from the Banach–Alaoglu theorem (Theorem 11.6); however, since this requires separability, we cannot apply it directly to $X^{**} \cong X$. We therefore proceed as follows: For given $\{x_n\}_{n \in \mathbb{N}} \subset B_X$, we set $U := \mathrm{cl}\,(\mathrm{lin}\,\{x_n : n \in \mathbb{N}\})$. This closed subspace is then again reflexive by Theorem 10.5 and separable by definition. Hence $U^{**} \cong U$ is also separable (since $\{J_U x_n\}_{n \in \mathbb{N}}$ is dense in $U^{**} = J_U(U)$), and therefore so is U^* by Corollary 8.7. Furthermore, the sequence $\{J_U x_n\}_{n \in \mathbb{N}} \subset U^{**}$ is bounded in U^{**} (since J_U is an isometry) and thus by Corollary 11.7 has a weakly-$*$ convergent subsequence $\{J_U x_{n_k}\}_{k \in \mathbb{N}}$ with $J_U x_{n_k} \rightharpoonup^* u^{**} \in U^{**}$. Since U is reflexive, there exists an $x := J_U^{-1} u^{**} \in U \subset X$. Since $u_n \in U$, it follows for all $x^* \in X^*$ that

$$\langle x^*, x_{n_k} \rangle_X = \langle x^*|_U, x_{n_k} \rangle_U = \langle J_U x_{n_k}, x^*|_U \rangle_{U^*}$$
$$\to \langle u^{**}, x^*|_U \rangle_{U^*} = \langle x^*|_U, x \rangle_U = \langle x^*, x \rangle_X,$$

and hence $x_{n_k} \rightharpoonup x$. $\qquad\square$

This also yields a "weak Bolzano–Weierstraß theorem".

> **Corollary 11.9**
>
> If X is a reflexive normed vector space, then every bounded sequence in X contains a weakly convergent subsequence.

We now have everything at hand to generalize the Weierstraß theorem (Corollary 2.9) to infinite-dimensional vector spaces.

---

[4]This terminology is not fully correct, either: the Eberlein–Šmulian theorem states that in Banach spaces the weak sequential compactness and the open cover compactness in the weak topology are equivalent. (Since weak topology is, in contrast to the weak-$*$ topology, not metrizable, this is a nontrivial statement.) Together with a theorem by Goldstine stating that the reflexivity of X is equivalent to the weak open cover compactness of B_X (see, e.g., [7, Theorem 4.2]), this yields the given special case (which—for separable spaces—was already proven by Banach as well). Again, one can often find this special case under the more general name.

Theorem 11.10 (Tonelli)
Let X be a reflexive normed vector space, let $U \subset X$ be nonempty, bounded, convex, and closed, and let $f : X \to \mathbb{R}$ be weakly lower semicontinuous. Then there exists an $\bar{x} \in U$ with

$$f(\bar{x}) = \min_{x \in U} f(x).$$

Proof. Since $\{f(x) : x \in U\} \subset \mathbb{R}$ is nonempty, there exists an $M := \inf_{x \in U} f(x) < \infty$ (where we have not yet excluded that $M = -\infty$, i.e., that f is not bounded from below). By the properties of the infimum, we can thus find a sequence $\{y_n\}_{n \in \mathbb{N}} \subset f(U) \subset \mathbb{R}$ such that $y_n \to M$, i.e., a sequence $\{x_n\}_{n \in \mathbb{N}} \subset U$ such that

$$f(x_n) = y_n \to M = \inf_{x \in U} f(x).$$

Such a sequence is called a *minimizing sequence*. Note that the convergence of $\{f(x_n)\}_{n \in \mathbb{N}}$ does not imply the convergence of $\{x_n\}_{n \in \mathbb{N}}$.

Now the boundedness of U implies in particular that the minimizing sequence is bounded; Corollary 11.9 thus yields a weakly convergent subsequence $\{x_{n_k}\}_{k \in \mathbb{N}}$ with limit $\bar{x} \in X$. Since U is convex and closed, $\{x_{n_k}\}_{k \in \mathbb{N}} \subset U$ implies that $\bar{x} \in U$ by Theorem 11.5. This limit is our candidate for a minimizer.

From the definition of the minimizing sequence, we also have $f(x_{n_k}) \to M$ as $k \to \infty$. Together with the weak lower semicontinuity of f and the definition of the infimum, we thus obtain that

$$\inf_{x \in U} f(x) \leq f(\bar{x}) \leq \liminf_{k \to \infty} f(x_{n_k}) = M = \inf_{x \in U} f(x).$$

Hence the infimum is attained at $\bar{x} \in U$, i.e., $-\infty < f(\bar{x}) = \min_{x \in U} f(x)$. □

If X is instead separable, we can similarly argue for X^* using weak-$*$ closedness and lower semicontinuity. These and similar results are fundamental (and referred to as *the direct method*) in the calculus of variations.

Problems

Problem 11.1 (*Weak convergence and operators*)
Let X and Y be Banach spaces, $\{x_n\}_{n \in \mathbb{N}} \subset X$ with $x_n \rightharpoonup x \in X$, and $T \in L(X, Y)$. Show that $Tx_n \rightharpoonup Tx$.

Problem 11.2 *(Weak cauchy sequences)*

Let X be a reflexive Banach space and $\{x_n\}_{n\in\mathbb{N}} \subset X$ a weak Cauchy sequence, i.e., $\{\langle x^*, x_n \rangle_X\}_{n\in\mathbb{N}} \subset \mathbb{F}$ is a Cauchy sequence for all $x^* \in X^*$. Show that $\{x_n\}_{n\in\mathbb{N}}$ converges weakly.

Problem 11.3 *(Weak convergence and dense subsets)*

Let X be a normed vector space. Show that a bounded sequence $\{x_n\}_{n\in\mathbb{N}}$ converges weakly to some $x \in X$ if and only if there is a subset $D \subset X^*$ such that $\mathrm{cl}(\mathrm{lin}\, D) = X^*$ and

$$\lim_{n\to\infty} \langle x^*, x_n \rangle_X = \langle x^*, x \rangle_X \quad \text{for all } x^* \in D.$$

Problem 11.4 *(Weak convergence)*

Determine whether the following sequences $\{x_n\}_{n\in\mathbb{N}} \subset \ell^2(\mathbb{F})$ converge weakly in $\ell^2(\mathbb{F})$ and if so, determine their limit:

(i) $x_n := a + e_n$ for given $a \in \ell^2(\mathbb{F})$ and unit vectors e_n in $\ell^2(\mathbb{F})$;
(ii) $x_n := n e_n$.

Problem 11.5 *(Weak-* convergence)*

Consider for $\{a_k\}_{k\in\mathbb{N}} \in \ell^\infty(\mathbb{F})$ the sequence $\{x_n\}_{n\in\mathbb{N}} \subset \ell^\infty(\mathbb{F})$ defined via

$$x_n := (0, \ldots, 0, a_{n+1}, a_{n+2}, \ldots),$$

i.e., $(x_n)_k = 0$ for all $k \neq n$. Show that $x_k \rightharpoonup^* 0$.

Problem 11.6 *(Weak-* convergence of the derivative)*

For all $\varepsilon > 0$ and $x \in C^1([-1, 1])$, set

$$f_\varepsilon(x) = \frac{1}{2\varepsilon}\left(x(\varepsilon) - x(-\varepsilon) \right) \quad \text{and} \quad f_0(x) = x'(0),$$

where $C^1([-1, 1])$ is endowed with the norm $\|x\|_{C^1} = \|x\|_\infty + \|x'\|_\infty$. Show that or give a counterexample for:

(i) $f_\varepsilon \in C^1([-1, 1])^*$ for all $\varepsilon \geq 0$;
(ii) $f_{\varepsilon_n} \rightharpoonup^* f_0$ for every null sequence $\{\varepsilon_n\}_{n\in\mathbb{N}} \subset [0, \infty)$;
(iii) $f_{\varepsilon_n} \to f_0$ for every null sequence $\{\varepsilon_n\}_{n\in\mathbb{N}} \subset [0, \infty)$.

Problem 11.7 *(Not weakly-* closed sets)*

Give an example of a set that is closed and convex but not weakly-* closed.
Hint: It has to be a subset of a nonreflexive dual space, and null spaces of linear operators are always closed and convex.

Problem 11.8 *(Mazur's theorem)*

Let X be a normed vector space and $\{x_n\}_{n\in\mathbb{N}} \subset X$ a sequence with $x_n \rightharpoonup x$. Show that there exists a sequence $\{y_n\}_{n\in\mathbb{N}} \subset X$ of convex combinations

$$y_n = \sum_{k=1}^{N_n} \lambda_{n,k} x_k \quad \text{with} \quad \sum_{k=1}^{N_n} \lambda_{n,k} = 1, \quad \lambda_{n,k} \geq 0, \quad N_n \in \mathbb{N},$$

with $y_n \to x$.

Hint: Consider the convex hull

$$\mathrm{co}\,\{x_n\}_{n\in\mathbb{N}} := \left\{ \sum_{k=1}^{N} \lambda_k x_k : \sum_{k=1}^{N} \lambda_k = 1,\ \lambda_k \geq 0,\ N \in \mathbb{N} \right\}.$$

Part IV

Compact Operators Between Banach Spaces

Compact Operators

12

We have seen that bounded sequences are guaranteed to contain a convergent subsequence only in finite-dimensional spaces; otherwise we can in general find only a *weakly* convergent subsequence. The reason for this is that in infinite-dimensional spaces, boundedness is not enough to ensure precompactness. One exception is the range of *compact* operators, since these inherit essential properties of finite-dimensional operators.

A linear mapping $T : X \to Y$ between the normed vector spaces X and Y is called *compact* if T maps bounded sets to relatively compact sets. Since relatively compact sets are always precompact and therefore bounded by Theorem 2.7, compact operators are always continuous by Definition 1.7 (i).

The linearity of T implies that as in Lemma 4.1, it suffices to consider the unit ball in the definition. Furthermore, we can use the equivalence of open cover and sequential compactness. This yields the following equivalent characterizations.

> **Lemma 12.1**
>
> *Let X and Y be Banach spaces and let $T : X \to Y$ be linear. Then the following properties are equivalent:*
>
> *(i) T is compact;*
> *(ii) $T(B_X)$ is precompact;*
> *(iii) if $\{x_n\}_{n\in\mathbb{N}} \subset X$ is bounded, then $\{Tx_n\}_{n\in\mathbb{N}} \subset Y$ contains a convergent subsequence.*

Proof. (iii) $\Rightarrow$ (ii): By Theorem 2.7, it suffices to show that every sequence in $T(B_X)$ contains a convergent subsequence. Let $\{y_n\}_{n\in\mathbb{N}} \subset T(B_X)$. Then there exists a sequence

© Springer Nature Switzerland AG 2020
C. Clason, *Introduction to Functional Analysis*, Compact Textbooks
in Mathematics, https://doi.org/10.1007/978-3-030-52784-6_12

$\{x_n\}_{n\in\mathbb{N}} \subset B_X$ with $y_n = Tx_n$. By assumption, $\{Tx_n\}_{n\in\mathbb{N}}$ contains a subsequence $\{Tx_{n_k}\}_{k\in\mathbb{N}}$ with $Tx_{n_k} \to y \in Y$, and hence $y_{n_k} = Tx_{n_k} \to y$.

(ii) $\Rightarrow$ (i): Due to the linearity of T, (ii) implies by a scaling argument that $T(B_R(0))$ is precompact for every $R > 0$. Furthermore, for every bounded $A \subset X$, by definition there exists an $R > 0$ such that $A \subset B_R(0)$ and hence $T(A) \subset T(B_R(0))$. Since subsets of precompact sets are again precompact, the claim follows from Theorem 2.7.

(i) $\Rightarrow$ (iii): If T is compact and $\{x_n\}_{n\in\mathbb{N}} \subset X$ is bounded, then $T(\{x_n\}_{n\in\mathbb{N}}) = \{Tx_n\}_{n\in\mathbb{N}}$ is relatively compact and therefore contains a convergent subsequence by Theorem 2.7.   □

The following alternative formulation of (iii) can be useful.

Lemma 12.2

Let X and Y be Banach spaces and let $T : X \to Y$ be linear. If X is reflexive, then the following properties are equivalent:

(i) T is compact;
(ii) T is completely continuous, i.e., $x_n \rightharpoonup x$ implies $Tx_n \to Tx$.

If X is not reflexive, then (i) $\Rightarrow$ (ii).

Proof. (i) $\Rightarrow$ (ii): Let $\{x_n\}_{n\in\mathbb{N}}$ a weakly convergent sequence with $x_n \rightharpoonup x \in X$. Then $\{x_n\}_{n\in\mathbb{N}}$ is bounded by Theorem 11.3, and hence Lemma 12.1 (iii) yields a convergent subsequence $\{Tx_{n_k}\}_{k\in\mathbb{N}}$ with $Tx_{n_k} \to y \in Y$. Now $x_n \rightharpoonup x$ implies that

$$\langle y^*, Tx_n\rangle_Y = \langle T^*y^*, x_n\rangle_X \to \langle T^*y^*, x\rangle_X = \langle y^*, Tx\rangle_Y \quad \text{for all } y^* \in Y^*$$

and hence that $Tx_n \rightharpoonup Tx$. The uniqueness of the limit then yields $y = Tx$. This shows that all convergent subsequences converge to the same limit, and hence $Tx_n \to Tx$ along the full sequence by a subsequence–subsequence argument.

(ii) $\Rightarrow$ (i): If X is reflexive, then Corollary 11.9 shows that every bounded sequence $\{x_n\}_{n\in\mathbb{N}}$ contains a weakly convergent subsequence $\{x_{n_k}\}_{k\in\mathbb{N}}$ with $x_{n_k} \rightharpoonup x \in X$. If now T is completely continuous, then this implies that $Tx_{n_k} \to Tx$ and hence that T is compact by Lemma 12.1 (iii).   □

Similarly to Corollary 4.6, we can show that the composition of compact operators is compact. Here it even suffices that only *one* of the operators is compact.

Lemma 12.3

Let X, Y, and Z be Banach spaces, $T \in L(X, Y)$, and $S \in L(Y, Z)$. If either T or S is compact, then $S \circ T$ is compact as well.

Proof. Let $\{x_n\}_{n\in\mathbb{N}} \in X$ be a bounded sequence. If T is compact, then $\{Tx_n\}_{n\in\mathbb{N}}$ contains a convergent subsequence, and the continuity of S further implies that $\{S(Tx_{n_k})\}_{k\in\mathbb{N}}$ is also convergent. Conversely, if S is compact, then the continuity of T implies that the sequence $\{Tx_n\}_{n\in\mathbb{N}}$ is bounded, which in turn implies that $\{S(Tx_n)\}_{n\in\mathbb{N}}$ contains a convergent subsequence. $\qquad\square$

In analogy to $L(X, Y)$, we now define the set

$$K(X, Y) := \{T : X \to Y : T \text{ is linear and compact}\}.$$

Since every compact operator is continuous, $K(X, Y) \subset L(X, Y)$. We even have the following

Theorem 12.4

Let X and Y be Banach spaces. Then $K(X, Y)$ is a closed subspace of $L(X, Y)$.

Proof. We first show that $K(X, Y)$ is a subspace. Let $S, T \in K(X, Y)$ and $\alpha \in \mathbb{F}$, and let $\{x_n\}_{n\in\mathbb{N}} \subset X$ be a bounded sequence. Since S and T are compact, there exists a convergent subsequence $\{Sx_n\}_{n\in\mathbb{N}_1}$ for some countably infinite set $\mathbb{N}_1 \subset \mathbb{N}$. Then the corresponding subsequence $\{x_n\}_{n\in\mathbb{N}_1}$ is also bounded, and hence there exists a convergent subsequence $\{Tx_n\}_{n\in\mathbb{N}_2}$ for another countably infinite subset $\mathbb{N}_2 \subset \mathbb{N}_1$. Since $\{Sx_n\}_{n\in\mathbb{N}_2}$ still converges, so does $\{\alpha Sx_n + Tx_n\}_{n\in\mathbb{N}_2}$. We have thus found a convergent subsequence of $\{(\alpha S+T)x_n\}_{n\in\mathbb{N}}$, implying that $\alpha S + T$ is compact.

Let now $\{T_n\}_{n\in\mathbb{N}} \subset K(X, Y)$ be a convergent sequence with $T_n \to T \in L(X, Y)$. We have to show that T is compact. To this end, let $\{x_m\}_{m\in\mathbb{N}}$ be bounded, i.e., $\{x_m\}_{m\in\mathbb{N}} \in B_R(0)$ for some $R > 0$. Using a diagonal sequence argument as in the proof of the Banach–Alaoglu theorem (Theorem 11.6), we can then construct a subsequence $\{x_{m_k}\}_{k\in\mathbb{N}}$ such that $\{T_n x_{m_k}\}_{k\in\mathbb{N}}$ converges for all $n \in \mathbb{N}$. We next show that $\{Tx_{m_k}\}_{k\in\mathbb{N}}$ is a Cauchy sequence. Since $T_n \to T$, there exists for every $\varepsilon > 0$ an $N \in \mathbb{N}$ with $\|T_N - T\|_{L(X,Y)} \leq \varepsilon$. Furthermore, $\{T_N x_{m_k}\}_{k\in\mathbb{N}}$ is convergent and hence a Cauchy sequence; we can thus find a $K > 0$ such that

$$\|T_N x_{m_j} - T_N x_{m_k}\|_Y \leq \varepsilon \qquad \text{for all } j, k \geq K.$$

Hence we have for all $j, k \geq K$ that

$$\|T x_{m_j} - T x_{m_k}\|_Y \leq \|T x_{m_j} - T_N x_{m_j}\|_Y + \|T_N x_{m_j} - T_N x_{m_k}\|_Y + \|T_N x_{m_k} - T x_{m_k}\|_Y$$

$$\leq \|T_N - T\|_{L(X,Y)} \|x_{m_j}\|_X + \|T_N x_{m_j} - T_N x_{m_k}\|_Y$$

$$+ \|T_N - T\|_{L(X,Y)} \|x_{m_k}\|_X$$

$$\leq \varepsilon R + \varepsilon + \varepsilon R = (2R + 1)\varepsilon.$$

This shows that $\{T x_{m_k}\}_{k \in \mathbb{N}} \subset Y$ is a Cauchy sequence and therefore convergent due to the completeness of Y. □

We now consider examples of (non)compact operators. Even the trivial example of the identity is interesting.

Lemma 12.5

Let X be a Banach space. Then $\mathrm{Id} : X \to X$ is compact if and only if X is finite-dimensional.

Proof. By Lemma 12.1 (ii), Id is compact if and only if $\mathrm{Id}(B_X) = B_X = \mathrm{cl}\, B_X$ is precompact and hence even compact. However, Theorem 3.11 shows that B_X is compact if and only if X is finite-dimensional. □

Corollary 12.6

Let X and Y be Banach spaces. If $T \in K(X, Y)$ is invertible, then X and Y are finite-dimensional.

Proof. Since $K(X, Y) \subset L(X, Y)$ by Theorem 12.4, T is even continuously invertible by Theorem 5.6. Lemma 12.3 then yields that $T \circ T^{-1} = \mathrm{Id}_Y$ as well as $T^{-1} \circ T = \mathrm{Id}_X$ is compact. By Lemma 12.5, this is possible only if X and Y are finite-dimensional. □

In other words, compact operators on infinite-dimensional spaces are never invertible!

The converse implication in Lemma 12.5 can be made more precise.

Lemma 12.7

Let X and Y be normed vector spaces and $T \in L(X, Y)$. If ran T is finite-dimensional, then T is compact.

Proof. For bounded $A \subset X$, it follows from the linearity and continuity of T that $T(A) \subset$ ran T is bounded as well. Hence cl $T(A)$ bounded and closed and therefore compact by the Heine–Borel theorem (Theorem 2.5). In particular, $T(A)$ is relatively compact, and hence T is a compact operator. $\square$

Since $K(X, Y)$ is closed, we obtain the following characterization of a useful class of compact operators.[1]

Corollary 12.8

Let X and Y be Banach spaces and $\{T_n\}_{n\in\mathbb{N}} \subset L(X, Y)$. If ran T is finite-dimensional for all $n \in \mathbb{N}$ and $T_n \to T \in L(X, Y)$, then T is compact.

Now for some concrete examples.

Example 12.9

(i) Let $X = Y = \ell^1(\mathbb{F})$ and

$$T : X \to Y, \qquad \{x_k\}_{k\in\mathbb{N}} \mapsto \left\{\tfrac{1}{k}x_k\right\}_{k\in\mathbb{N}}.$$

To show compactness, we define for each $n \in \mathbb{N}$ the operator

$$T_n : X \to Y, \qquad \{x_k\}_{k\in\mathbb{N}} \mapsto \{y_k\}_{k\in\mathbb{N}}, \qquad y_k := \begin{cases} \tfrac{1}{k}x_k & \text{if } k \leq n, \\ 0 & \text{if } k > n. \end{cases}$$

It is straightforward to verify that T_n is bounded and has finite-dimensional range (with $\dim(\text{ran } T_n) = n$). Let now $x \in \ell^1(\mathbb{F})$ be arbitrary. Estimating

$$\|Tx - T_nx\|_{\ell^1} = \sum_{k=n+1}^{\infty} \tfrac{1}{k}|x_k| \leq \frac{1}{n+1} \sum_{k=n+1}^{\infty} |x_k| \leq \frac{1}{n+1}\|x\|_{\ell^1},$$

we have that $\|T - T_n\|_{L(X,Y)} \leq \frac{1}{n+1} \to 0$, and hence T is compact by Corollary 12.8.

(ii) Let $X = Y = C([0, 1])$ and define the *integration operator*

$$T : X \to Y, \qquad x \mapsto \int_0^t x(s)\, ds.$$

----

[1] The converse implication holds only under additional assumptions on X: there are compact operators that are not the limit of a sequence of continuous operators with finite-dimensional range; see [10].

To show compactness, we prove that $T(B_X) \subset C([0, 1])$ is precompact using the Arzelà–Ascoli theorem (Theorem 2.11). Estimating

$$\|Tx\|_\infty = \sup_{t \in [0,1]} \left| \int_0^t x(s)\,ds \right| \leq \sup_{t \in [0,1]} |t| \|x\|_\infty = \|x\|_\infty,$$

we obtain that $T(B_X)$ is bounded pointwise, i.e., $\sup_{t \in [0,1]} |Tx(t)| \leq 1$ for all $x \in B_X$. Furthermore, we have for all $\varepsilon > 0$, $x \in B_X$, and $t_1, t_2 \in [0, 1]$ with $|t_1 - t_2| \leq \varepsilon$ that

$$|[Tx](t_1) - [Tx](t_2)| = \left| \int_{t_1}^{t_2} x(s)\,ds \right| \leq |t_1 - t_2| \|x\|_\infty \leq \varepsilon,$$

which yields equicontinuity and therefore the precompactness of $T(B_X)$. Hence T is compact.

These examples illustrate that compact operators are "smoothing": Tx and Ty are always more similar than x and y. It thus stands to reason that such operators are not (continuously and hence in Banach spaces not at all) invertible, since otherwise their inverses would magnify small differences correspondingly. This similarity of course depends on the norms. For example, considering the integration operator

$$T : C([0, 1]) \to C_0^1([0, 1]) := \left\{ f \in C^1([0, 1]) : f(0) = 0 \right\},$$

then T is invertible if we endow $C_0^1([0, 1])$ with the norm $\|x\|_{C^1} = \|x'\|_\infty + \|x\|_\infty$: in this case, clearly $T^{-1} = D$ for the derivative operator $D : C^1([0, 1]) \to C([0, 1])$, which was already shown in Example 4.4 (iv) to be continuous between these spaces. Hence T cannot be compact by Corollary 12.6 since $C([0, 1])$ is infinite-dimensional. This shows that both invertibility and compactness depend crucially on the norms, where it is impossible to have both at the same time. (In this sense, simultaneous continuity of T and T^{-1} is the best possible compromise.)

We close this chapter by studying the adjoint of compact operators.

Theorem 12.10 (Schauder)
Let X and Y be Banach spaces and $T \in L(X, Y)$. Then T^* is compact if and only if T is compact.

Proof. Let T be compact and let $\{y_n^*\}_{n \in \mathbb{N}} \subset Y^*$ be a bounded sequence, i.e., $\{y_n^*\}_{n \in \mathbb{N}} \subset B_R(0)$ for some $R > 0$. We show that $\{T^* y_n^*\}_{n \in \mathbb{N}} \subset X^*$ contains a convergent subsequence by first constructing a convergent subsequence of preimages y_n^*, which the continuous operator T^* then maps to the desired convergent subsequence. For this, we use the assumption that $T(B_X)$ is relatively compact, i.e., that $K := \mathrm{cl}\, T(B_X) \subset Y$ is compact.

We first show that the sequence $\{y_n^* |_K\}_{n \in \mathbb{N}} \subset K^*$ contains a convergent subsequence. Since every continuous linear functional in K^* is in particular a continuous function on K and hence an element of $C(K)$, we can apply the Arzelà–Ascoli theorem (Theorem 2.11). We first have that

$$\|y_n^* |_K\|_{C(K)} = \sup_{y \in K} |\langle y_n^*, y \rangle_Y| \leq \sup_{y \in K} \|y_n^*\|_{Y^*} \|y\|_Y \leq R \sup_{y \in K} \|y\|_Y < \infty$$

since K is compact and therefore bounded. Hence $\{y_n^* |_K\}_{n \in \mathbb{N}} \subset C(K)$ is pointwise bounded in particular. Furthermore,

$$|\langle y_n^*, y_1 \rangle_Y - \langle y_n^*, y_2 \rangle_Y| \leq \|y_n^*\|_{Y^*} \|y_1 - y_2\|_Y \leq R \|y_1 - y_2\|_Y \quad \text{for all } y_1, y_2 \in K,$$

and thus $\{y_n^* |_K\}_{n \in \mathbb{N}}$ is equicontinuous. Hence $\{y_n^* |_K\}_{n \in \mathbb{N}}$ is precompact and therefore contains a subsequence $\{y_{n_k}^* |_K\}_{k \in \mathbb{N}}$ converging with respect to the supremum norm. We now show that the corresponding subsequence $\{T^* y_{n_k}^*\}_{k \in \mathbb{N}} \subset X^*$ is a Cauchy sequence. For all $k, l \in \mathbb{N}$,

$$\|T^* y_{n_k}^* - T^* y_{n_l}^*\|_{X^*} = \sup_{x \in B_X} |\langle y_{n_k}^* - y_{n_l}^*, Tx \rangle_Y| = \sup_{y \in K} |\langle y_{n_k}^* - y_{n_l}^*, y \rangle_Y|$$

$$= \|y_{n_k}^* |_K - y_{n_l}^* |_K\|_{C(K)}$$

since $T(B_X)$ is dense in K by definition. Since $\{y_{n_k}^* |_K\}_{k \in \mathbb{N}}$ is a Cauchy sequence, this implies that $\{T^* y_{n_k}^*\}_{k \in \mathbb{N}} \subset X^*$ is a Cauchy sequence as well and thus convergent by the completeness of Y. This shows that T^* is compact.

Conversely, if T^* is compact, then T^{**} is compact by the first part of the proof. Lemma 10.3 then yields that $J_Y \circ T = T^{**} \circ J_X$. The canonical embedding $J_Y : Y \to Y^{**}$ is always injective and continuous; it also has a closed range since Y is a Banach space. Hence J_Y is continuously invertible on $\operatorname{ran}(T^{**} \circ J_X) \subset \operatorname{ran} J_Y$. The compactness of $T = J_Y^{-1} \circ T^{**} \circ J_X$ now follows from Lemma 12.3. $\qquad \square$

Problems

Problem 12.1 *(Compact operators on ℓ^p)*
Let $1 \leq p < \infty$ and $z \in \ell^\infty(\mathbb{F})$. Let further $T_z : \ell^p(\mathbb{F}) \to \ell^p(\mathbb{F})$ be defined via $[T_z x]_k := z_k x_k$ for all $x \in \ell^p$ and $k \in \mathbb{N}$. Show that T_z is compact if and only if z is a null sequence.

Problem 12.2 *(Compact operators on $C([0, 1])$)*
Show or give a counterexample for the compactness of $S : C([0, 1]) \to C([0, 1])$, defined via $[Sx](t) = tx(t)$ for all $x \in C([0, 1])$ and $t \in [0, 1]$.

Problem 12.3 *(Compact operators on ℓ^2)*

Let $a_{jk} \in \mathbb{R}$, $j, k \in \mathbb{N}$, be given with $\sum_{j=1}^{\infty} \sum_{k=1}^{\infty} |a_{jk}|^2 < \infty$. Show that

$$K : \ell^2(\mathbb{R}) \to \ell^2(\mathbb{R}), \qquad x \mapsto \left(\sum_{k=1}^{\infty} a_{1k} x_k, \sum_{k=1}^{\infty} a_{2k} x_k, \dots \right),$$

is compact.

Problem 12.4 *(Fredholm integral operator)*

Let $k : [0, 1] \times [0, 1] \to \mathbb{R}$ be a continuous function. Show that

(i) for all $x \in C([0, 1])$, the function $T_k x$, defined via

$$[T_k x](s) = \int_0^1 k(s, t) x(t) dt \quad \text{for all } s \in [0, 1],$$

is continuous, i.e., the operator $T_k : C([0, 1]) \to C([0, 1])$ is well-defined;

(ii) $T_k : C([0, 1]) \to C([0, 1])$ is a compact linear operator.

Problem 12.5 *(Pointwise limit of compact operators)*

Show by a counterexample that the *pointwise* limit of a sequence of compact operators need not be compact.

Hint: Consider a suitable sequence in $K(\ell^2(\mathbb{R}), \ell^2(\mathbb{R}))$ with finite-dimensional range.

Problem 12.6 *(Ehrling's lemma)*

Let X, Y, and Z be Banach spaces, let $T \in K(X, Y)$ be compact, and let $S \in L(Y, Z)$ be injective. Show that for every $\varepsilon > 0$ there exists a $C_\varepsilon > 0$ such that

$$\|Tx\|_Y \le \varepsilon \|x\|_X + C_\varepsilon \|TSx\|_X \qquad \text{for all } x \in X.$$

Hint: Attempt a proof by contradiction.

The Fredholm Alternative

13

We have seen that compact operators on infinite-dimensional spaces are never invertible since their range is "too small" in a certain sense. The situation changes if we add the identity—this results in a *compact perturbation of the identity*, and since the identity is invertible, chances are good that the sum is still invertible. Such operators are of interest since operator equations of the form $\lambda x = Tx$ for some (given) $\lambda \in \mathbb{F} \setminus \{0\}$ occur in many applications, e.g., for eigenvalue problems (see the following chapter) or fixed-point equations (for $\lambda^{-1}T$). Since T is compact if and only if $\lambda^{-1}T$ is compact, we can restrict ourselves in the following to the case $\lambda = 1$. For the remainder of this chapter, let X be a Banach space, $\mathrm{Id} : X \to X$ the identity, $T \in K(X) := K(X, X)$, and $S := \mathrm{Id} - T$ our compact perturbation of the identity. We start by showing some fundamental properties.

Lemma 13.1

If $T \in K(X)$, then $\ker(\mathrm{Id} - T)$ is finite-dimensional.

Proof. For all $x \in \ker(\mathrm{Id} - T) = \ker S$, we have by definition that $\mathrm{Id}\, x = Tx$, i.e., $\mathrm{Id}\,|_{\ker S} = T\,|_{\ker S}$. Since T is compact, $\mathrm{Id} : \ker S \to \ker S$ is therefore compact as well, and hence $\ker S$ must be finite-dimensional by Lemma 12.5. $\qquad\square$

Lemma 13.2

If $T \in K(X)$, then $\mathrm{ran}(\mathrm{Id} - T)$ is closed.

© Springer Nature Switzerland AG 2020
C. Clason, *Introduction to Functional Analysis*, Compact Textbooks in Mathematics, https://doi.org/10.1007/978-3-030-52784-6_13

Proof. We apply Lemma 9.7. Assume to the contrary that ran $S = \text{ran}(\text{Id} - T)$ is not closed. Then (9.1) does not hold either, i.e., we can find a sequence $\{x_n\}_{n \in \mathbb{N}} \in X$ with $\|[x_n]\|_{X/\ker S} = 1$ and $\|S x_n\|_X \to 0$. As in the proof of Lemma 9.7, we can furthermore choose x_n such that $\|x_n\|_X \le 2$.

Since $\{x_n\}_{n \in \mathbb{N}}$ is bounded and T is compact, there exists a convergent subsequence of $\{T x_n\}_{n \in \mathbb{N}}$ with $T x_{n_k} \to z \in X$. It then follows from $S x_n \to 0$ that

$$x_{n_k} = (\text{Id} - T) x_{n_k} + T x_{n_k} = S x_{n_k} + T x_{n_k} \to 0 + z = z.$$

Since the quotient mapping and the norm are continuous, we obtain that

$$\|[z]\|_{X/\ker S} = \lim_{k \to \infty} \|[x_{n_k}]\|_{X/\ker S} = 1. \tag{13.1}$$

On the other hand, since S is continuous, we have that $S z = \lim_{k \to \infty} S x_{n_k} = 0$, i.e., $z \in \ker S$ and hence $\|[z]\|_{X/\ker S} = 0$, in contradiction to (13.1). $\qquad \square$

By Schauder's theorem (Theorem 12.10), compactness of T implies compactness of T^*, and hence $\ker(\text{Id} - T^*) = \ker S^*$ is finite-dimensional as well. In turn, it follows from Theorems 7.3 and 9.6 (ii) that

$$X/\text{ran}\, S \cong (\text{ran}\, S)^{\perp} = \ker S^*$$

is finite-dimensional; in particular, $\dim(X/\text{ran}\, S) = \dim(\ker S^*)$. (This number is called the *codimension* $\text{codim}(\text{ran}\, S) := \dim(X/\text{ran}\, S)$ of ran S.) An operator $S \in L(X, Y)$ for which $\ker S$ and $X/\text{ran}\, S$ is called a *Fredholm operator*; compact perturbations of the identity are the most prominent but not the only examples. The number

$$\text{ind}\,(S) := \dim(\ker S) - \dim(X/\text{ran}\, S)$$

is then called the *index* of S. If $\text{ind}\,(S) = 0$, then $\dim(\ker S) = \dim(X/\text{ran}\, S)$, which generalizes the rank–nullity theorem from linear algebra. In particular, in this case S is injective if and only if S is surjective. Otherwise the index indicates how many dimensions are missing for one or the other: if the index is negative, S is not surjective; if it is positive, S is not injective. We will not study this "quantitative" Fredholm theory further and only show that for $S = \text{Id} - T$, injectivity and surjectivity are equivalent (which is a slightly weaker statement than $\text{ind}\,(S) = 0$).

Theorem 13.3

Let $T \in K(X)$ and $S = \text{Id} - T$. Then S is injective if and only if S is surjective.

Proof. Assume to the contrary that S is injective but not surjective, i.e., that there exists an $x \in X \setminus \text{ran}\, S$. We then show that T cannot be compact by constructing a sequence $\{x_n\}_{n \in \mathbb{N}}$

that does not lead to a convergent subsequence $\{Tx_{n_k}\}_{k\in\mathbb{N}}$. To this end, we consider the subspaces $U_n := \operatorname{ran} S^n$ for $S^n := S \circ \cdots \circ S$. By the binomial formula,

$$S^n = (\operatorname{Id} - T)^n = \operatorname{Id} + \sum_{k=1}^{n} \binom{n}{k}(-T)^k =: \operatorname{Id} + \tilde{T}.$$

Since the composition of compact operators is compact and $K(X)$ is a subspace, $\tilde{T}$ is compact, and hence U_n is closed by Lemma 13.2. Furthermore, $U_m \subset U_n$ for all $m > n$ by definition of the range. We now show that this inclusion is strict. First, $x \notin \operatorname{ran} S$ by definition implies that $S^n x \in U_n$. Assuming that $S^n x \in U_{n+1}$ as well would yield an $y \in X$ with $S^{n+1} y = S^n x$, i.e.,

$$0 = S^{n+1} y - S^n x = S^n(Sy - x)$$

and hence $Sy = x$ by injectivity of S, in contradiction to $x \notin \operatorname{ran} S$. This shows that U_{n+1} is a proper subspace of U_n, and Riesz's lemma (Lemma 3.10) therefore yields an $x_n \in U_n$ such that $\|x_n\|_X = 1$ and $\|u - x_n\|_X \geq \frac{1}{2}$ for all $u \in U_{n+1}$. This implies that

$$\|T(x_n - x_m)\|_X = \|S(x_n - x_m) + x_m - x_n\|_X \geq \frac{1}{2} \qquad \text{for all } m > n$$

since the U_n are nested and hence $S(x_n - x_m) + x_m \in U_{n+1}$ for all $m > n$. It follows that $\{Tx_n\}_{n\in\mathbb{N}}$ cannot contain a Cauchy sequence even though $\{x_n\}_{n\in\mathbb{N}}$ is bounded, which contradicts the compactness of T.

Conversely, let S be surjective. Then $S^* = \operatorname{Id}_{X^*} + T^*$ is injective by Theorem 9.6 (i) and therefore also surjective by the above. The injectivity of S then follows from Theorem 9.6 (iii).   □

We immediately obtain the following result on the solvability of the equation $\lambda x = Tx$, which we will need for proving the main result of the next chapter.

Corollary 13.4 (Fredholm alternative)
Let $T \in K(X)$ and $\lambda \in \mathbb{F} \setminus \{0\}$. Then one and only one *of the following properties holds:*

(i) The homogeneous equation

$$\lambda x - Tx = 0$$

has only the trivial solution $x = 0$, and the inhomogeneous equation

$$\lambda x - Tx = y$$

has a unique solution for every $y \in X$.

(ii) The homogeneous equation has a nontrivial solution $x \neq 0$, and the inhomogeneous equation has a solution if and only if $y \in (\ker(\lambda \operatorname{Id} - T^*))_\perp$.

Proof. Since ran S is closed by Lemma 13.1, we have that ran $S = $ cl ran $S = (\ker S^*)_\perp$ by Theorem 9.6. The claim now follows from Theorem 13.3. □

Problems

Problem 13.1 *(Invertible plus compact is Fredholm)*
Let $S \in L(X)$ be invertible and $T \in K(X)$ compact. Show that $S + T$ is a Fredholm operator.

Problem 13.2 *(Compact operators are not Fredholm)*
Let X be an infinite-dimensional Banach space and $T \in K(X)$. Show that T is not a Fredholm operator.
Hint: Use Theorem 6.4.

Problem 13.3 *(Fredholm operators on ℓ^p)*
Show that the shift operators

$$S_+ : (x_0, x_1, x_2, x_3, \dots) \mapsto (0, x_0, x_1, x_2, \dots) \quad \text{and}$$

$$S_- : (x_0, x_1, x_2, x_3, \dots) \mapsto (x_1, x_2, x_3, x_4, \dots)$$

are Fredholm operators on $\ell^p(\mathbb{F})$ for $1 \le p \le \infty$ with ind $S_+ = 1$ and ind $S_- = -1$.

Problem 13.4 *(Derivative as Fredholm operator)*
Show that the derivative operator

$$T : C^1([0, 1]) \to C([0, 1]), \qquad x \mapsto x',$$

is a Fredholm operator and determine ind T.

The Spectrum

14

An important tool in linear algebra is the study of eigenvalues and eigenvectors of a linear operator. Recall that $\lambda \in \mathbb{F}$ is called an *eigenvalue* of the linear operator $T : X \to X$ if there exists an $x \in X \setminus \{0\}$ with $Tx = \lambda x$, i.e., if $\lambda \operatorname{Id} - T$ is not injective. In finite-dimensional spaces, this is the case if and only if $\lambda \operatorname{Id} - T$ is not surjective, and this fact is heavily exploited in linear algebra. In infinite-dimensional spaces, however, we have to distinguish the cases of noninjectivity and nonsurjectivity.

We thus define for a Banach space X and $T \in L(X) := L(X, X)$

(i) the *point spectrum*

$$\sigma_p(T) := \{\lambda \in \mathbb{F} : \lambda \operatorname{Id} - T \text{ is not injective}\},$$

(ii) the *continuous spectrum*

$$\sigma_c(T) := \{\lambda \in \mathbb{F} : \lambda \operatorname{Id} - T \text{ is injective but not surjective, with dense range}\},$$

(iii) the *residual spectrum*

$$\sigma_r(T) := \{\lambda \in \mathbb{F} : \lambda \operatorname{Id} - T \text{ is injective but not surjective, without dense range}\},$$

as well as the *spectrum*

$$\sigma(T) := \sigma_p(T) \cup \sigma_c(T) \cup \sigma_r(T) = \{\lambda \in \mathbb{F} : \lambda \operatorname{Id} - T \text{ is not bijective}\},$$

where the union is clearly disjoint. Usually, $\sigma_p(T)$ is a union of discrete points, $\sigma_c(T)$ is a union of intervals, and $\sigma_r(T)$ is empty. Note that only $\lambda \in \sigma_p(T)$ is an

© Springer Nature Switzerland AG 2020
C. Clason, *Introduction to Functional Analysis*, Compact Textbooks in Mathematics, https://doi.org/10.1007/978-3-030-52784-6_14

eigenvalue of T. In this case, an $x \neq 0$ with $Tx = \lambda x$ is called an *eigenvector*; the closed subspace $\ker(\lambda \operatorname{Id} - T)$ is called an *eigenspace*. A fundamental property of eigenspaces is that they are *invariant* under T, i.e., $x \in \ker(\lambda \operatorname{Id} - T)$ implies that $Tx = \lambda x \in \ker(\lambda \operatorname{Id} - T)$. (For this reason, eigenspaces are also called *invariant subspaces*.)

Example 14.1 Consider the right-shift operator

$$S_+ : \ell^p(\mathbb{F}) \to \ell^p(\mathbb{F}), \qquad (x_1, x_2, , x_3, \dots) \mapsto (0, x_1, x_2, \dots).$$

Then

$$(\lambda \operatorname{Id} - S_+)x = (\lambda x_1, \lambda x_2 - x_1, \lambda x_3 - x_2, \dots)$$

is always injective, and hence $\sigma_p(S_+) = \emptyset$. However, it is possible to show that this operator is not surjective if and only if $|\lambda| \leq 1$, which implies that $\sigma(S_+) = B_{\mathbb{F}}$. Furthermore, since $\operatorname{ran} S_+ = \{x \in \ell^p(\mathbb{F}) : x_1 = 0\}$ is a closed subspace, we have, e.g., $0 \in \sigma_r(S_+)$.

It is often more convenient to study in place of $\sigma(T)$ its complement

$$\rho(T) := \mathbb{F} \setminus \sigma(T) = \{\lambda \in \mathbb{F} : \lambda \operatorname{Id} - T \text{ is bijective}\},$$

which is called the *resolvent set* of T. If $\lambda \in \rho(T)$, we correspondingly call

$$T_\lambda := (\lambda \operatorname{Id} - T)^{-1} \in L(X)$$

the *resolvent* of T, where the continuity of T_λ follows from the bounded inverse theorem (Theorem 5.6). We can obtain further information on the resolvent using the following

Lemma 14.2 (Neumann series)
Let X be a Banach space and $T \in L(X)$ with $\|T\|_{L(X)} < 1$. Then $\operatorname{Id} - T$ is bijective and satisfies

$$(\operatorname{Id} - T)^{-1} = \sum_{k=0}^{\infty} T^k.$$

Proof. If $\|T\|_{L(X)} < 1$, we obtain from Corollary 4.6 that

$$\sum_{k=0}^{\infty} \|T^k\|_{L(X)} \le \sum_{k=0}^{\infty} \|T\|_{L(X)}^k < \infty.$$

Hence the Neumann series converges (absolutely) to some $S \in L(X)$. Consider now the sequence $\{S_n\}_{n \in \mathbb{N}}$ with $S_n = \sum_{k=0}^{n} T^k$. Then

$$(\mathrm{Id} - T)S_n = \sum_{k=0}^{n} T^k - \sum_{k=0}^{n} T^{k+1} = T^0 - T^{n+1} = \mathrm{Id} - T^{n+1}.$$

It follows from $\|T\|_{L(X)} < 1$ that $\|T^n\|_{L(X)} \le \|T\|_{L(X)}^n \to 0$ as $n \to \infty$. The continuity of $\mathrm{Id} - T$ now yields

$$(\mathrm{Id} - T)S = \lim_{n \to \infty} (\mathrm{Id} - T)S_n = \mathrm{Id}.$$

Analogously, one shows that $S(\mathrm{Id} - T) = \mathrm{Id}$ and hence $S = (\mathrm{Id} - T)^{-1}$, which in particular implies that $\mathrm{Id} - T$ is invertible. $\qquad\square$

We now use the Neumann series to expand T_λ locally into a power series.

Lemma 14.3

Let X be a Banach space, $T \in L(X)$, and $\lambda_0 \in \rho(T)$. Then

$$T_\lambda = \sum_{k=0}^{\infty} (\lambda_0 - \lambda)^k T_{\lambda_0}^{k+1} \qquad \text{for all } |\lambda - \lambda_0| < \|T_{\lambda_0}\|_{L(X)}^{-1}.$$

In particular, $\rho(T)$ is open.

Proof. We start by writing

$$\lambda \, \mathrm{Id} - T = (\lambda_0 \, \mathrm{Id} - T) - (\lambda_0 - \lambda) \, \mathrm{Id} = (\lambda_0 \, \mathrm{Id} - T)(\mathrm{Id} - (\lambda_0 - \lambda)(\lambda_0 \, \mathrm{Id} - T)^{-1})$$

$$=: (\lambda_0 \, \mathrm{Id} - T)(\mathrm{Id} - \tilde{T}). \tag{14.1}$$

Since $\lambda_0 \in \rho(T)$, the operator $\lambda_0 \, \mathrm{Id} + T$ is invertible. Furthermore, if $\|\tilde{T}\|_{L(X)} = |\lambda - \lambda_0| \, \|T_{\lambda_0}\|_{L(X)} < 1$, then $\mathrm{Id} - \tilde{T}$ is invertible by Lemma 14.2. Hence $\lambda \, \mathrm{Id} - T$ is invertible as well. In particular, if λ is sufficiently close to $\lambda_0 \in \rho(T)$, then $\lambda_0 \in \rho(T)$ as well, which shows that $\rho(T)$ is open.

We can thus invert both sides of (14.1) to obtain

$$T_\lambda = (\lambda \operatorname{Id} - T)^{-1} = \left(\operatorname{Id} - (\lambda_0 - \lambda)(\lambda_0 \operatorname{Id} - T)^{-1} \right)^{-1} (\lambda_0 \operatorname{Id} - T)^{-1}$$

$$= \left(\sum_{k=0}^{\infty} \left((\lambda_0 - \lambda) T_{\lambda_0} \right)^k \right) T_{\lambda_0} = \sum_{k=0}^{\infty} (\lambda_0 - \lambda)^k T_{\lambda_0}^{k+1}. \qquad \square$$

By passing to the complement, this implies useful properties of the spectrum.

Theorem 14.4

Let X be a Banach space and $T \in L(X)$. Then

(i) $\sigma(T)$ is compact;

(ii) $|\lambda| \leq \|T\|_{L(X)}$ for all $\lambda \in \sigma(T)$;

(iii) if $\mathbb{F} = \mathbb{C}$ and $X \neq \{0\}$, then $\sigma(T)$ is nonempty.

Proof. Assume that $T \neq 0$ (otherwise $\sigma(T) = \{0\}$ and hence the claims hold trivially). Consider first $\lambda \in \mathbb{F}$ with $|\lambda| > \|T\|_{L(X)}$. Then Lemma 14.2 implies that $\operatorname{Id} - \lambda^{-1} T$ is invertible; since $\lambda \neq 0$, it follows that

$$\lambda \left(\operatorname{Id} - \lambda^{-1} T \right) = \lambda \operatorname{Id} - T \qquad (14.2)$$

is invertible as well. Hence $\lambda \in \rho(T)$, and passing to the complement yields $\sigma(T) \subset B_{\|T\|_{L(X)}}$ and therefore (ii). Furthermore, this implies that the spectrum is bounded; it is also closed by Lemma 14.3 and hence compact by the Heine–Borel theorem (Theorem 2.5), which shows (i).

For (iii), let $\mathbb{F} = \mathbb{C}$ and $X \neq \{0\}$. Since we have to prove a statement about complex numbers, we apply techniques from complex analysis. To this end, we consider for an arbitrary continuous linear functional $\xi \in L(X)^*$ the complex function

$$f : \rho(T) \to \mathbb{C}, \qquad \lambda \mapsto \langle \xi, T_\lambda \rangle_{L(X)}.$$

Let now $\lambda_0 \in \rho(T)$ be arbitrary. It then follows from the continuity of ξ together with Lemma 14.3 that

$$f(\lambda) = \langle \xi, T_\lambda \rangle_{L(X)} = \sum_{k=0}^{\infty} (\lambda_0 - \lambda)^k \langle \xi, T_{\lambda_0}^{k+1} \rangle_{L(X)} \quad \text{for all } \lambda \in U_{\|T_{\lambda_0}\|_{L(X)}^{-1}}(\lambda_0).$$

This implies that for every point inside an open disk, f can be expanded into a power series. Hence f is holomorphic (i.e., complex differentiable) and therefore in particular continuous (see, e.g., [20, Theorem 10.6]).

Assume now to the contrary that $\sigma(T) = \emptyset$ and hence $\rho(T) = \mathbb{C}$. We then show that f is bounded on $\mathbb{C}$ and therefore constant by Liouville's theorem (see, e.g., [20, Theorem 10.23]). On the one hand, the continuous function f is bounded on the compact set $\{\lambda : |\lambda| \leq 2\|T\|_{L(X)}\}$ by the Weierstraß theorem (Corollary 2.9). For $|\lambda| > 2\|T\|_{L(X)}$, on the other hand, Lemma 14.2 and (14.2) imply that

$$T_\lambda = (\lambda \operatorname{Id} - T)^{-1} = \lambda^{-1} \sum_{k=0}^{\infty} (\lambda^{-1} T)^k.$$

Since $\|\lambda^{-1} T\|_{L(X)} < \frac{1}{2}$, we thus have

$$|f(\lambda)| = |\langle \xi, T_\lambda \rangle_{L(X)}| \leq \frac{1}{|\lambda|} \|\xi\|_{L(X)^*} \sum_{k=0}^{\infty} \|\lambda^{-1} T\|_{L(X)}^k \leq \frac{2}{|\lambda|} \|\xi\|_{L(X)^*}$$

$$\leq \|T\|_{L(X)}^{-1} \|\xi\|_{L(X)^*}. \tag{14.3}$$

Hence f is bounded on all of $\mathbb{C}$ and therefore constant; taking $|\lambda| \to \infty$ on the right-hand side of the first line of (14.3) shows that this constant has to be zero.

By definition of f, this implies that $\langle \xi, T_\lambda \rangle_{L(X)} = f(\lambda) = 0$ for all $\lambda \in \mathbb{C}$. Since $\xi \in L(X)^*$ was arbitrary, it follows from Corollary 8.4 that $T_\lambda = 0$, in contradiction to $T_\lambda = (\lambda \operatorname{Id} - T)^{-1}$ and $X \neq \{0\}$. □

The estimate $|\lambda| \leq \|T\|_{L(X)}$ can be strengthened. To this end, we define the *spectral radius*

$$r(T) := \sup_{\lambda \in \sigma(T)} |\lambda|.$$

Then Theorem 14.4 implies that $r(T) \leq \|T\|_{L(X)}$ and that the supremum is attained if $\sigma(T)$ is nonempty. As in linear algebra, we can derive an expression for the spectral radius in terms of the norm of powers of T. We start with the spectrum of polynomials of T, where we define for a (complex) polynomial $p(z) := \sum_{k=0}^{n} a_k z^k$ of degree at most n and $T \in L(X)$ the linear operator

$$p(T) := \sum_{k=0}^{n} a_k T^k,$$

where $T^k := T \circ \cdots \circ T$ denotes the k times repeated composition and $T^0 := \operatorname{Id}$. Clearly, $p(T) : X \to X$ is linear as well as bounded by Corollary 4.6. The spectrum of $p(T)$ then has the following catchy representation.

Lemma 14.5 (Spectral polynomial theorem)

Let X be a Banach space over $\mathbb{F} = \mathbb{C}$, $T \in L(X)$, and p a polynomial. Then

$$\sigma(p(T)) = p(\sigma(T)) := \{p(\lambda) : \lambda \in \sigma(T)\}.$$

Proof. For a constant polynomial $p(z) = a_0 \in \mathbb{C}$, the operator $p(T) = a_0 \,\mathrm{Id}$ clearly has only the eigenvalue a_0. We can thus assume that p has degree $n \geq 1$.

Let now $\lambda \in \sigma(p(T))$ and consider the polynomial $q(z) := p(z) - \lambda$, which also has degree n and hence has n complex roots $\lambda_1, \ldots, \lambda_n$ by the fundamental theorem of algebra. We can thus write $q(z) = \gamma \sum_{k=1}^{n}(z - \lambda_j)$ for some $\gamma \neq 0$ and therefore

$$p(T) - \lambda \,\mathrm{Id} = q(T) = \gamma \sum_{k=1}^{n}(T - \lambda_j \,\mathrm{Id}).$$

By assumption, $p(T) - \lambda \,\mathrm{Id}$ is not bijective, which implies that at least one of the factors $T - \lambda_j \,\mathrm{Id}$ cannot be bijective. It follows that $\lambda_j \in \sigma(T)$ as well as $\lambda = p(\lambda_j)$ (since λ_j is a root of q), i.e, $\lambda \in p(\sigma(T))$.

Conversely, let $\lambda \in \sigma(T)$ be arbitrary and consider the polynomial $q(z) := p(z) - p(\lambda)$. Since λ is clearly a root of q, we can write $q(z) = (z - \lambda)r(z)$ for some polynomial r of degree at most $n - 1$ and therefore

$$p(T) - p(\lambda) \,\mathrm{Id} = q(T) = (T - \lambda \,\mathrm{Id})r(T).$$

By assumption, $T - \lambda \,\mathrm{Id}$ is not bijective, which implies that $p(T) - p(\lambda) \,\mathrm{Id}$ cannot be bijective either, i.e., $p(\lambda) \in \sigma(p(T))$. $\qquad\square$

We can now prove the promised expression for the spectral radius.

Theorem 14.6

Let X be a Banach space over $\mathbb{F} = \mathbb{C}$ and $T \in L(X)$. Then

$$r(T) = \lim_{n \to \infty} \|T^n\|_{L(X)}^{1/n} = \inf_{n \in \mathbb{N}} \|T^n\|_{L(X)}^{1/n}.$$

Proof. We first show that $r(T) \leq \inf_{n \in \mathbb{N}} \|T^n\|_{L(X)}^{1/n}$. To this end, let $\lambda \in \sigma(T)$ be arbitrary; such a λ exists since $\sigma(T)$ is nonempty by Theorem 14.4. The spectral polynomial theorem (Lemma 14.5) then yields $\lambda^n \in \sigma(T^n)$ for all $n \in \mathbb{N}$, and Theorem 14.4 implies that $|\lambda^n| \leq \|T^n\|_{L(X)}$ and hence that $|\lambda| \leq \|T^n\|_{L(X)}^{1/n}$. Taking the supremum over all $\lambda \in \sigma(T)$ and the infimum over all $n \in \mathbb{N}$ then yields the claimed lower bound for $r(T)$.

Next, we show that $r(T) \geq \limsup_{n\to\infty} \|T^n\|_{L(X)}^{1/n}$ by continuing the proof of Theorem 14.4. Recall that we have shown there that for every $\xi \in L(X)^*$, the function $f : \rho(T) \to \mathbb{C}, \lambda \mapsto \langle \xi, T_\lambda \rangle_{L(X)}$ is holomorphic and for $|\lambda| > \|T\|_{L(X)}$ can be written as

$$f(\lambda) = \sum_{k=0}^{\infty} \lambda^{-k-1} \langle \xi, T^k \rangle_{L(X)}; \tag{14.4}$$

see (14.3). We now use the fact that as a holomorphic function, f can be expanded on the unbounded annulus $K_r := \{\lambda : |\lambda| > r(T)\} \subset \rho(T)$ into a Laurent series of the form $\sum_{k=-\infty}^{\infty} a_k \lambda^k$; see, e.g., [16, Theorem 2.1]. Since (14.4) already provides a series in the requisite form for $\lambda \in \{\lambda : |\lambda| > \|T\|_{L(X)}\} \subset K_r$, the expression (14.4) even has to be valid for all $\lambda \in K_r$. In particular, this implies that $\{\langle \xi, \lambda^{-k-1} T^k \rangle_{L(X)}\}_{k\in\mathbb{N}}$ is a null sequence. Since $\xi \in L(X)^*$ was arbitrary, it follows that $\lambda^{-k-1} T^k \rightharpoonup 0$ and hence that $\{\lambda^{-k-1} T^k\}_{k\in\mathbb{N}} \subset L(X)$ is bounded by Theorem 11.3. We can thus find a constant $C > 0$ such that

$$\|T^k\|_{L(X)}^{1/k} \leq (|\lambda|^{k+1} C)^{1/k} = |\lambda|(C|\lambda|)^{1/k} \qquad \text{for all } |\lambda| > r(T).$$

Since the right-hand side converges as $k \to \infty$, we obtain that

$$\limsup_{k\to\infty} \|T^k\|_{L(X)}^{1/n} \leq \lim_{n\to\infty} |\lambda|(C|\lambda|)^{1/k} = |\lambda| \qquad \text{for all } |\lambda| > r(T).$$

Passing to the limit $|\lambda| \to r(T)$ then yields the claimed upper bound for $r(T)$.

We thus deduce that

$$r(T) \leq \inf_{n\in\mathbb{N}} \|T^n\|_{L(X)}^{1/n} \leq \liminf_{n\to\infty} \|T^n\|_{L(X)}^{1/n} \leq \limsup_{n\to\infty} \|T^n\|_{L(X)}^{1/n} \leq r(T),$$

since the second and third inequalities hold for arbitrary sequences. Hence all inequalities hold with equality, and we obtain the desired expression. □

Although a linear operator (with the trivial exception $T = 0$) always has a nonempty spectrum, it need not contain any eigenvalues. The situation improves for compact operators.

Theorem 14.7 (Little spectral theorem)
Let $T \in K(X)$ be compact. Then

(i) $\sigma(T) \subset \sigma_p(T) \cup \{0\}$;
(ii) $\ker(\lambda \operatorname{Id} - T)$ is finite-dimensional for all $\lambda \in \sigma_p(T) \setminus \{0\}$;
(iii) $\sigma_p(T)$ is finite or countably infinite and can have only 0 as an accumulation point.

Proof. (i): By definition, $\lambda \operatorname{Id} - T$ is injective for all $\lambda \notin \sigma_p(T) \cup \{0\}$, and hence $\operatorname{Id} - \lambda^{-1} T$ is injective as well. Furthermore, since T is compact, $\lambda^{-1} T$ is also compact for all $\lambda \neq 0$. It follows from Theorem 13.3 that $\operatorname{Id} - \lambda^{-1} T$ is surjective, and hence $\lambda \operatorname{Id} - T$ is surjective as well. Again by definition, this yields $\lambda \in \rho(T) = \mathbb{F} \setminus \sigma(T)$.

(ii): For compact T and $\lambda \neq 0$, Lemma 13.1 implies that $\ker(\lambda \operatorname{Id} - T) = \ker(\operatorname{Id} - \lambda^{-1} T)$ is finite-dimensional.

(iii): Assume that $\sigma_p(T)$ is infinite (otherwise there is nothing to show) and that $\{\lambda_n\}_{n\in\mathbb{N}} \subset \sigma_p(T)$ is a sequence of eigenvalues. By passing to a subsequence if necessary, we can assume that these eigenvalues are all distinct. We can then choose for every λ_n an eigenvector $x_n \in X$ with $\|x_n\|_X = 1$. The same argument as in linear algebra shows that eigenvectors corresponding to different eigenvalues are linearly independent (since linear combinations by definition consist of only finitely many terms). The subspace $X_n := \operatorname{lin}\{x_1, \ldots, x_n\}$ thus has $\dim X_n = n$. We now use Riesz's lemma (Lemma 3.10) to find for X_{n-1} and $x_n \notin X_{n-1}$ a $v_n \in X_n$ with $\|v_n\|_X = 1$ and

$$\|v_n - x\|_X \geq \frac{1}{2} \qquad \text{for all } x \in X_{n-1}.$$

(For $n = 1$, we set $X_0 := \emptyset$ and can thus take $v_1 = x_1$.) Since the X_n are nested, we can write $v_n = \alpha_n x_n + \tilde{v}_{n-1}$ for some $\alpha_n \in \mathbb{F}$ and $\tilde{v}_{n-1} \in X_{n-1}$. Furthermore, since X_{n-1} is spanned by eigenvectors and is thus an invariant subspace, $\tilde{v}_m \in X_{n-1}$ and therefore also $T\tilde{v}_m \in X_{n-1}$ for all $m < n$. But this implies that

$$(\lambda_n \operatorname{Id} - T)v_n = 0 + (\lambda_n \operatorname{Id} - T)\tilde{v}_{n-1} \in X_{n-1}.$$

It follows for all $n > m \in \mathbb{N}$ that

$$\left\| T\left(\frac{v_n}{\lambda_n}\right) - T\left(\frac{v_m}{\lambda_m}\right) \right\|_X = \left\| v_n - \lambda_n^{-1}\left[(\lambda_n \operatorname{Id} - T)v_n - \tfrac{\lambda_n}{\lambda_m} T v_m \right] \right\|_X \geq \frac{1}{2}$$

by choice of v_n, since the vector in brackets is an element of X_{n-1}. Hence $\{\lambda_n^{-1} T v_n\}_{n\in\mathbb{N}}$ cannot contain a convergent subsequence. Since T is compact, this is possible only if *every* subsequence of $\{\lambda_n^{-1} v_n\}_{n\in\mathbb{N}}$ is unbounded. We thus have along the full sequence

$$\frac{1}{|\lambda_n|} = \frac{\|v_n\|_X}{|\lambda_n|} = \|\lambda_n^{-1} v_n\|_X \to \infty,$$

i.e., $\{\lambda_n\}_{n\in\mathbb{N}}$ is a null sequence. Since $\{\lambda_n\}_{n\in\mathbb{N}} \subset \sigma_p(T)$ was arbitrary, $\sigma_p(T)$ can accumulate only at 0. In particular, for every $\varepsilon > 0$ the set $\sigma_\varepsilon(T) := \{\lambda \in \sigma_p(T) : |\lambda| > \varepsilon\}$ can contain only finitely many elements, and hence $\sigma_p(T) \subset \bigcup_{n\in\mathbb{N}} \sigma_{\frac{1}{n}}(T) \cup \{0\}$ is at most countable. $\qquad\square$

Problems

Problem 14.1 *(Spectrum of compact operators)*
Let X be an infinite-dimensional Banach space and let $T \in K(X)$ be compact. Show that $0 \in \sigma(T)$.

Problem 14.2 *(Spectrum of adjoint operators)*
Let X be a Banach space and $T \in L(X, X)$. Show that $\sigma(T^*) = \sigma(T)$.

Problem 14.3 *(Spectrum of the shift operators)*

(i) Determine the spectrum of the left-shift operator

$$S_- : \ell^1(\mathbb{F}) \to \ell^1(\mathbb{F}), \quad (x_1, x_2, x_3, \ldots) \mapsto (x_2, x_3, x_4, \ldots),$$

as well as of its adjoint $S_-^* : \ell^\infty(\mathbb{F}) \to \ell^\infty(\mathbb{F})$.
(ii) Which elements of the spectrum are eigenvalues?
(iii) What changes when we consider $S_- : \ell^2(\mathbb{F}) \to \ell^2(\mathbb{F})$?

Problem 14.4 *(Spectrum of the integration operator)*
Show that the integration operator

$$T : C([0, 1]) \to C([0, 1]), \quad [Tx](t) = \int_0^t x(s)ds,$$

satisfies $\sigma(T) = \sigma_r(T) = \{0\}$.

Problem 14.5 *(Spectrum of the multiplication operator)*

(i) Determine for given $h \in C([0, 1])$ the spectrum of the *multiplication operator*

$$T_h : C([0, 1]) \to C([0, 1]), \quad [Th](t) = f(t)h(t).$$

(ii) Give necessary and sufficient conditions for an element of the spectrum to be an eigenvalue.
(iii) Give necessary and sufficient conditions for T_h to be compact.

Problem 14.6 *(Spectrum of perturbations of the identity)*
Let $T \in L(X)$ with $\|T\|_{L(X)} \in \sigma(T)$. Show that

$$\| \mathrm{Id} + T \|_{L(X)} = 1 + \|T\|_{L(X)}.$$

Part V

Hilbert Spaces

Inner Products and Orthogonality

15

Further results on linear operators are possible in Hilbert spaces, where the algebraic and topological structures of a normed vector space are joined by an additional, geometric, structure: the inner product. We will see in the next chapters that this allows a complete characterization without resorting to dual spaces, leading to a generalization of the structure theory of linear algebra from Euclidean vector spaces to infinite-dimensional (Hilbert) spaces.

Definition 15.1

Let X be a vector space over $\mathbb{F} \in \{\mathbb{R}, \mathbb{C}\}$. An *inner product* on X is a mapping $(\cdot, \cdot)_X : X \times X \to \mathbb{F}$ with the following properties:

(i) $(\lambda x_1 + x_2, y)_X = \lambda (x_1, y)_X + (x_2, x)_X$ for all $x_1, x_2, y \in X$ and $\lambda \in \mathbb{F}$;

(ii) $(x, y)_X = \overline{(y, x)_X}$ for all $x, y \in \mathbb{F}$;

(iii) $(x, x)_X \geq 0$ for all $x \in X$, with $(x, x)_X = 0$ if and only if $x = 0 \in X$.

In this case, the pair $(X, (\cdot, \cdot)_X)$ is called an *inner product space* or a *pre-Hilbert space*. If the inner product is obvious from the context, we simply write X for the inner product space.

Properties (i) and (ii) immediately imply that

$$(x, \lambda y_1 + y_2)_X = \overline{\lambda} (x, y_1)_X + (x, y_2)_X \quad \text{for all } x, y_1, y_2 \in X \text{ and } \lambda \in \mathbb{F},$$

i.e., the inner product is *sesquilinear* ("one-and-a-half times linear"); note that this definition is not consistent in the literature (i.e., property (i) may instead require

© Springer Nature Switzerland AG 2020
C. Clason, *Introduction to Functional Analysis*, Compact Textbooks
in Mathematics, https://doi.org/10.1007/978-3-030-52784-6_15

linearity in the *second* argument), but this does not essentially affect the following results. In particular, since $\lambda + \bar{\lambda} = 2\,\mathrm{Re}\,\lambda$ we have the *binomial expansion*

$$(x+y, x+y)_X = (x, x)_X + 2\,\mathrm{Re}\,(x, y)_X + (y, y)_X \qquad \text{for all } x, y \in X. \qquad (15.1)$$

Furthermore, property (ii) also implies that $(x, x)_X \in \mathbb{R}$ for all $x \in X$ even for $\mathbb{F} = \mathbb{C}$, and hence property (iii) is meaningful. Despite the formal similarity, one should not confuse the (sesquilinear, symmetric) inner product in X with the (bilinear, not symmetric) duality pairing between X and X^*.

A fundamental property of the inner product is the following inequality.

Theorem 15.2   (Cauchy–Schwarz inequality)

Let $(X, (\cdot, \cdot)_X)$ be an inner product space. Then

$$|(x, y)_X| \le \sqrt{(x, x)_X}\sqrt{(y, y)_X} \qquad \text{for all } x, y \in X.$$

Proof. For all $x, y \in X$ and $\lambda \in \mathbb{F}$, the binomial expansion (15.1) together with $\lambda\bar{\lambda} = |\lambda|^2 \in \mathbb{R}$ implies that

$$0 \le (x+\lambda y, x+\lambda y)_X = (x, x)_X + 2\,\mathrm{Re}\,\bar{\lambda}\,(x, y)_X + |\lambda|^2\,(y, y)_X\,.$$

Let now $y \ne 0$ (otherwise the claim holds trivially). Then we obtain for $\lambda = -\frac{(x, y)_X}{(y, y)_X}$ that

$$0 \le (x, x)_X - 2\frac{|(x, y)_X|^2}{(y, y)_X} + \frac{|(x, y)_X|^2}{(y, y)_X} = (x, x)_X - \frac{|(x, y)_X|^2}{(y, y)_X},$$

which after rearranging and taking the square root yields the claim.   □

This inequality is an essential result since it guarantees that the new structure is compatible with the previously introduced structures, just as the norm is compatible with the metric (induced by that norm).

Theorem 15.3

Let $(X, (\cdot, \cdot)_X)$ be an inner product space. Then

$$\|x\|_X := \sqrt{(x, x)_X}$$

defines a norm on X. If $(X, \|\cdot\|_X)$ is complete, it is called a Hilbert space.

Proof. The norm axioms follow directly from those of the inner product: if $\|x\|_X = 0$, then $(x, x)_X = 0$ and hence $x = 0$. Furthermore, all $\lambda \in \mathbb{F}$ and $x \in X$ satisfy

$$\|\lambda x\|_X^2 = (\lambda x, \lambda x)_X = \lambda \bar{\lambda} (x, x)_X = |\lambda|^2 \|x\|_X^2$$

and hence the positive homogeneity. The triangle inequality follows from the Cauchy–Schwarz inequality: for all $x, y \in X$, the definition of the norm together with $\mathrm{Re}\, \lambda \leq |\lambda|$ implies that

$$\|x + y\|_X^2 = \|x\|_X^2 + 2\, \mathrm{Re}\, (x, y)_X + \|y\|_X^2$$

$$\leq \|x\|_X^2 + 2\|x\|_X \|y\|_X + \|y\|_X^2 = (\|x\|_X + \|y\|_X)^2. \qquad \square$$

Every inner product space therefore corresponds in a canonical way to a normed vector space, which we will not distinguish in the following. Hence if we speak of norms, neighborhoods, or convergent sequences in Hilbert space, these are always meant with respect to the induced norm.[1] In particular, the Cauchy–Schwarz inequality from Theorem 15.2 immediately implies that the inner product is continuous (with respect to the induced norm) in each component.

Conversely, the inner product can be expressed through the induced norm: for $\mathbb{F} = \mathbb{R}$,

$$(x, y)_X = \frac{1}{4} \left(\|x + y\|_X^2 - \|x - y\|_X^2 \right); \qquad (15.2)$$

for $\mathbb{F} = \mathbb{C}$,

$$(x, y)_X = \frac{1}{4} \left(\|x + y\|_X^2 - \|x - y\|_X^2 + i\|x + iy\|_X^2 - i\|x - iy\|_X^2 \right). \qquad (15.3)$$

These *polarization identities* can be verified easily using the binomial expansion (15.1). In fact, they uniquely characterize norms that are induced by inner products.

Theorem 15.4 (Parallelogram identity)

Let $(X, \| \cdot \|_X)$ be a normed vector space. Then $\| \cdot \|_X$ is induced by an inner product if and only if

$$\|x + y\|_X^2 + \|x - y\|_X^2 = 2 \left(\|x\|_X^2 + \|y\|_X^2 \right) \qquad \textit{for all } x, y \in X.$$

---

[1] However, there are in fact situations in which it is useful *not* to take the induced norm; see the remark at the end of Chap. 16.

Proof. If X is an inner product space, the parallelogram identity is a direct consequence of (15.1). Conversely, one can use the parallelogram identity to verify that (15.2) and (15.3) define inner products; the somewhat tedious calculations can be found in, e.g., [13]. □

The parallelogram identity can also be used to verify the following examples.

Example 15.5 The following spaces, each endowed with its canonical norm, are Hilbert spaces:

   (i) $\mathbb{F}^n$ with the inner product $(x, y)_{\mathbb{F}^n} := \sum_{k=1}^{n} x_n \overline{y_n}$;
   (ii) $\ell^2(\mathbb{F})$ with the inner product $(x, y)_{\ell^2} := \sum_{k=1}^{\infty} x_n \overline{y_n}$;
   (iii) $L^2(\Omega)$ with the inner product $(x, y)_{L^2} := \int_{\Omega} x(t) \overline{y(t)} \, dt$.

The following space is an inner product space but not a Hilbert space:

(iv) $c_c(\mathbb{F}) \subset \ell^2(\mathbb{F})$ with the inner product from (ii).

The following spaces are *not* inner product spaces:

   (v) $\ell^p(\mathbb{F})$ and $L^p(\Omega)$ for $p \neq 2$;
  (vi) $C(K)$ for a compact set $K \neq \{0\}$.

Just as the norm generalizes the geometric notion of "length", the inner product generalizes the notion of "angle"—where the right angle is as usual of special importance. Correspondingly, we call two vectors $x, y \in X$ *orthogonal* if $(x, y)_X = 0$. In this case, the binomial expansion (15.1) becomes the *Pythagorean theorem*

$$\|x + y\|_X^2 = \|x\|_X^2 + \|y\|_X^2.$$

Furthermore, we define for an arbitrary set $A \subset X$ the *orthogonal complement*

$$A^{\perp} := \left\{ x \in X : (x, y)_X = 0 \text{ for all } y \in A \right\}.$$

Again, despite the formal similarity, the orthogonal complement (as a subset of X) should not be confused with the annihilator (as a subset of X^*). However, the exact same arguments show that $A^{\perp}$ is always closed and that $\mathrm{cl}\, A \subset (A^{\perp})^{\perp}$.

We now come to a central result of Hilbert space theory, which guarantees a *unique* projection onto convex sets. This result crucially depends on both completeness and the parallelogram identity. (Compare also to Tonelli's theorem, Theorem 11.10.)

> **Theorem 15.6 (Projection theorem)**
> *Let X be a Hilbert space and let $C \subset X$ be nonempty, convex, and closed. Then for every $x \in X$, there exists a unique $z \in C$ such that*
>
> $$\|z - x\|_X = \inf_{y \in C} \|y - x\|_X.$$
>
> *The mapping $P_C : X \to C$, $x \mapsto z$, is called the (metric) projection onto C.*

Proof. We first show existence using the completeness of X. To this end, let $x \in X$ be given and set $d := \inf_{y \in C} \|y - x\|_X$; this infimum is finite since C is nonempty and the norm is nonnegative. By the properties of the infimum, we can thus find a sequence $\{y_n\}_{n \in \mathbb{N}} \subset C$ with $\|y_n - x\|_X \to d$. We now show that $\{y_n\}_{n \in \mathbb{N}}$ is a Cauchy sequence. First, the parallelogram identity implies for all $n, m \in \mathbb{N}$ that

$$2 \left(\|y_n - x\|_X^2 + \|y_m - x\|_X^2 \right) = \|(y_n + y_m) - 2x\|_X^2 + \|y_n - y_m\|_X^2$$

and hence that

$$\|y_n - y_m\|_X^2 = 2 \left(\|y_n - x\|_X^2 + \|y_m - x\|_X^2 \right) - 4 \left\| \tfrac{y_n + y_m}{2} - x \right\|_X^2. \tag{15.4}$$

Since C is convex, $y_n, y_m \in C$ implies that $\frac{1}{2} y_n + \frac{1}{2} y_m \in C$ as well, and thus the definition of d yields

$$0 \le \|y_n - y_m\|_X^2 \le 2 \left(\|y_n - x\|_X^2 + \|y_m - x\|_X^2 \right) - 4d^2.$$

By definition of the sequence $\{y_n\}_{n \in \mathbb{N}}$, the right-hand side tends to zero as $n, m \to \infty$. Hence $\{y_n\}_{n \in \mathbb{N}}$ is a Cauchy sequence and therefore converges to some $z \in X$ due to the completeness of X. Since C is closed, we even have $z \in C$. The continuity of the norm then yields

$$\|z - x\|_X = \lim_{n \to \infty} \|y_n - x\|_X = d = \inf_{y \in C} \|y - x\|_X.$$

To show uniqueness, let $z, \tilde{z} \in C$ be given with $\|z - x\|_X = d = \|\tilde{z} - x\|_X$. By the convexity of C, we then also have $\frac{1}{2} z + \frac{1}{2} \tilde{z} \in C$, which as in (15.4) implies that

$$\|z - \tilde{z}\|_X^2 = 2 \left(\|z - x\|_X^2 + \|\tilde{z} - x\|_X^2 \right) - 4 \left\| \tfrac{z + \tilde{z}}{2} - x \right\|_X^2 = 4d^2 - 4 \left\| \tfrac{z + \tilde{z}}{2} - x \right\|_X^2 \le 0,$$

i.e., $z = \tilde{z}$. $\qquad\square$

The metric projection can also be characterized via the inner product.

Lemma 15.7
Let X be a Hilbert space and let $C \subset X$ be nonempty, convex, and closed. Then the following are equivalent for all $x, z \in X$:

(i) $z = P_C(x)$;
(ii) $z \in C$ and $\mathrm{Re}\,(z - x, y - z)_X \geq 0$ for all $y \in C$.

Proof. (ii) $\Rightarrow$ (i): The binomial expansion (15.1) together with (ii) implies for all $y \in C$ that

$$\|y - x\|_X^2 = \|(z - x) + (y - z)\|_X^2 = \|z - x\|_X^2 + 2\,\mathrm{Re}\,(z - x, y - z)_X + \|y - z\|_X^2$$
$$\geq \|z - x\|_X^2,$$

i.e., $z = P_C(x)$ by the projection theorem (Theorem 15.6).

(i) $\Rightarrow$ (ii): Let $z = P_C(x) \in C$ and let $y \in C$ be arbitrary. Since C is convex, we also have $y_t := (1 - t)z + ty \in C$ for all $t \in (0, 1)$. It follows from the projection theorem that

$$\|z - x\|_X^2 \leq \|y_t - x\|_X^2 = \|(z - x) + t(y - z)\|_X^2$$
$$= \|z - x\|_X^2 + 2t\,\mathrm{Re}\,(z - x, y - z)_X + t^2\|y - z\|_X^2.$$

Subtracting $\|z - x\|_X^2$ and dividing by $2t > 0$ then implies

$$0 \leq \mathrm{Re}\,(z - x, y - z)_X + \frac{t}{2}\|y - z\|_X^2,$$

and passing to the limit $t \to 0$ yields (ii). □

An important special case is that in which C is a closed subspace.

Corollary 15.8
Let X be a Hilbert space and $U \subset X$ a closed subspace. Then the following are equivalent for all $x, z \in X$:
(i) $z = P_U(x)$;
(ii) $z \in U$ and $(z - x, u)_X = 0$ for all $u \in U$.

Proof. Since subspaces are always convex, we can apply Lemma 15.7; it remains only to show that (ii$'$) is equivalent to (ii$'$).

(ii$'$) $\Rightarrow$ (ii): Let $y \in U$ be arbitrary. Then $z \in U$ implies that $u := y - z \in U$ as well, and hence (ii$'$) is a special case of (ii).

$(ii) \Rightarrow (ii')$: Let $u \in U$ be arbitrary. Then $y := u + z \in U$ as well, and (ii) implies that

$$\mathrm{Re}\,(z - x, u)_X = \mathrm{Re}\,(z - x, y - z)_X \geq 0 \qquad \text{for all } u \in U.$$

Inserting $-u \in U$ shows that even $\mathrm{Re}\,(z - x, u)_X = 0$ holds for all $u \in U$. Similarly, inserting $-iu \in U$ and using the sesquilinearity of the inner product together with $\mathrm{Re}(-ix) = \mathrm{Im}(x)$ shows that $\mathrm{Im}\,(z - x, u)_X = 0$, which yields (ii'). $\qquad\square$

In this case, the projection has an additional useful property; it is then called an *orthogonal projection*.

Theorem 15.9

Let X be a Hilbert space and $U \subset X$ a closed subspace. Then
   (i) $P_U \in L(X, X)$;
   (ii) $\|P_U\|_{L(X,X)} = 1$ if $U \neq \{0\}$;
   (iii) $\ker P_U = U^\perp$;
   (iv) $P_{U^\perp} = \mathrm{Id} - P_U$.

Proof. All four claims follow from Corollary 15.8: $z = P_U(x)$ if and only if $z \in U$ and $z - x \in U^\perp$.

(i) Since $U^\perp$ is a subspace, we have for all $\lambda_1, \lambda_2 \in \mathbb{F}$, $x_1, x_2 \in X$, and $z_1 = P_U(x_1)$, $z_2 = P_U(x_2)$ that

$$(\lambda_1 z_1 + \lambda_2 z_2) - (\lambda_1 x_1 + \lambda_2 x_2) = \lambda_1(x_1 - z_1) + \lambda_2(x_2 - z_1) \in U^\perp,$$

i.e., $\lambda_1 P_U(x_1) + \lambda_2 P_U(x_2) = P_U(\lambda_1 x_1 + \lambda_2 x_2)$. This shows the linearity of P_U. Furthermore, $P_U(x) - x \in U^\perp$ for all $x \in X$ implies that

$$\|x\|_X^2 = \|x - z + z\|_X^2 = \|x - z\|_X^2 + 2\,\mathrm{Re}\,(x - z, z)_X + \|z\|_X^2 \geq \|z\|_X^2, \qquad (15.5)$$

i.e., $\|P_U(x)\|_X = \|z\|_X \leq \|x\|_X$. This shows the continuity of P_U.
(ii) First, it follows from (15.5) that $\|P_U\|_{L(X,X)} \leq 1$. If now $x \in U \setminus \{0\}$, then $z := x \in U$ and $z - x = 0 \in U^\perp$. Hence $P_U(x) = x$, which implies that $\|P_U\|_{L(X,X)} = 1$.
(iii) We have $P_U(x) = 0 \in U$ if and only if $0 - x = -x \in U^\perp$. Since U is a subspace, the latter holds if and only if $x \in U^\perp$.
(iv) We have to show that every $x \in X$ and $z := x - P_U(x)$ satisfy $z \in U^\perp$ and $z - x \in (U^\perp)^\perp$. The former follows from the fact that $P_U(x) - x \in U^\perp$, the latter from

$$z - x = (x - P_U(x)) - x = -P_U(x) \in U \subset (U^\perp)^\perp. \qquad\square$$

In particular, Theorem 15.9 (iv) implies that if U is a closed subspace, then every $x \in X$ can be uniquely written as $x = u + u_\perp$ for a $u \in U$ and a $u_\perp \in U^\perp$.

This allows showing an analogous result to Corollary 8.8—without the use of a Hahn–Banach theorem.

Corollary 15.10
Let X be a Hilbert space and $U \subset X$ a subspace. Then $(U^{\perp})^{\perp} = \mathrm{cl}\, U$.

Proof. First, arguing exactly as in the proof of Corollary 8.8 shows that $\mathrm{cl}\, U \subset (U^{\perp})^{\perp}$. Conversely, let $x \in (U^{\perp})^{\perp}$. We now consider the closed subspace $V := \mathrm{cl}\, U$; Theorem 15.9 (iv) thus yields a $v \in V$ and a $v^{\perp} \in V^{\perp}$ such that $x = v + v^{\perp}$. It follows from $U \subset \mathrm{cl}\, U$ and the definition of the orthogonal complement that $V^{\perp} \subset U^{\perp}$ and hence that $v^{\perp} \in U^{\perp}$.[2] On the other hand, we have $v^{\perp} = x - v \in (U^{\perp})^{\perp}$ since $v \in V = \mathrm{cl}\, U \subset (U^{\perp})^{\perp}$ and $x \in (U^{\perp})^{\perp}$, which implies that

$$\|v^{\perp}\|_X^2 = \left(v^{\perp}, v^{\perp}\right)_X = \left(x - v, v^{\perp}\right)_X = 0.$$

Hence $v^{\perp} = 0$, and therefore $x = v \in V = \mathrm{cl}\, U$.   □

We obtain from this a criterion for the invertibility of linear operators on Hilbert spaces, which (in a slightly more complicated form) is one of the cornerstones of the modern theory of partial differential equations.

Theorem 15.11 (Lax–Milgram)
Let X be a Hilbert space and $T \in L(X, X)$. If there exists a $\gamma > 0$ such that

$$|(Tx, x)_X| \geq \gamma \|x\|_X^2 \qquad \text{for all } x \in X, \tag{15.6}$$

then T is invertible with $\|T^{-1}\|_{L(X,X)} \leq \gamma^{-1}$.

Proof. First, (15.6) together with the Cauchy–Schwarz inequality yields

$$\|Tx\|_X \geq \gamma \|x\|_X \qquad \text{for all } x \in X, \tag{15.7}$$

and thus by Corollary 9.8 with $C = \gamma^{-1}$ both the injectivity of T and the closedness of $\mathrm{ran}\, T$. Let now $x \in (\mathrm{ran}\, T)^{\perp}$ be given. Then $(Tx, x)_X = 0$, and (15.6) implies that $x = 0$,

---

[2] A limiting argument as in the proof of Corollary 8.8 even shows that $V^{\perp} = U^{\perp}$.

i.e., $(\operatorname{ran} T)^{\perp} = \{0\}$. It thus follows from Corollary 15.10 that

$$\operatorname{ran} T = ((\operatorname{ran} T)^{\perp})^{\perp} = \{0\}^{\perp} = X,$$

and hence T is surjective and therefore invertible. For all $y \in X$, we can thus insert $x := T^{-1}y$ in (15.7) to obtain

$$\|T^{-1}y\|_X \le \frac{1}{\gamma}\|TT^{-1}y\|_X = \frac{1}{\gamma}\|y\|_X. \qquad \square$$

In finite-dimensional vector spaces, the orthogonal projection onto a subspace can be computed explicitly using basis vectors; this can be generalized to Hilbert spaces. We call a subset $S \subset X$ of an inner product space an *orthonormal system* if all $u, v \in S$ satisfy

$$(u, v)_X = \begin{cases} 0 & \text{if } u \ne v, \\ 1 & \text{if } u = v. \end{cases}$$

(If only the first condition holds, i.e., if $\|u\|_X \ne 1$ for some $u \in S$, then S is called an *orthogonal system*.) We postpone the question about the existence of orthonormal systems and first study their properties. We start with the projection onto *finite-dimensional* subspaces.

Lemma 15.12

Let X be an inner product space, $S \subset X$ an orthonormal system, and $U = \operatorname{lin}\{e_1, \dots, e_n\}$ for some $e_1, \dots, e_n \in S$. Then

$$P_U x = \sum_{k=1}^{n} (x, e_k)_X \, e_k \qquad \text{for all } x \in X.$$

Proof. For given $x \in X$ we set $z := \sum_{k=1}^{n} (x, e_k)_X \, e_k$. Then $z \in U$ by construction, and the orthogonality of the e_j for all $1 \le j \le n$ implies that

$$\big(x - z, e_j\big)_X = \big(x, e_j\big)_X - \sum_{k=1}^{n} (x, e_k)_X \big(e_k, e_j\big)_X = \big(x, e_j\big)_X - \big(x, e_j\big)_X = 0.$$

Hence $(x - z, u)_X = 0$ for all $u \in U$, and therefore $z = P_U x$ by Corollary 15.8. $\qquad \square$

Corollary 15.13 (Bessel's inequality)
Let X be an inner product space and let $\{e_1, \dots, e_n\} \subset X$ be a finite orthonormal system. Then

$$\sum_{k=1}^{n} |(x, e_k)_X|^2 \le \|x\|_X^2 \qquad \text{for all } x \in X.$$

Proof. Set $U := \lin\{e_1, \dots, e_n\}$. Then Lemma 15.12 yields for all $x \in X$ that

$$\|P_U x\|_X^2 = \left(\sum_{k=1}^{n} (x, e_k)_X \, e_k, \, \sum_{j=1}^{n} (x, e_j)_X \, e_j \right)_X$$

$$= \sum_{k=1}^{n} \sum_{j=1}^{n} (x, e_k)_X \, \overline{(x, e_j)_X} \, (e_k, e_j)_X = \sum_{k=1}^{n} |(x, e_k)_X|^2. \qquad (15.8)$$

The claim now follows from $\|P_U x\|_X \le \|x\|_X$ by Theorem 15.9 (ii). $\square$

The question is now whether this also works for (countably) infinite-dimensional subspaces, i.e., whether we can pass to the limit $n \to \infty$ in Bessel's inequality.

Theorem 15.14
Let X be a Hilbert space and $S = \{e_n : n \in \mathbb{N}\}$ an orthonormal system. Then the following properties are equivalent:

(i) $\lin S$ *is dense in X;*

(ii) $x = \displaystyle\sum_{k=1}^{\infty} (x, e_k)_X \, e_k$ *for all $x \in X$;*

(iii) Parseval's identity,

$$\|x\|_X^2 = \sum_{k=1}^{\infty} |(x, e_k)_X|^2 \qquad \text{for all } x \in X.$$

If these properties hold, S is called an orthonormal basis.

Proof. (i) $\Rightarrow$ *(ii):* Set $U_m := \lin\{e_1, \dots, e_m\}$ and $P_m := P_{U_m}$. Let now $x \in X$ be given and let $\{x_n\}_{n \in \mathbb{N}} \subset \lin S$ be a sequence with $x_n \to x$. Since linear combinations by definition are always finite, we can find for every $x_n \in \lin S$ an $m_n \in \mathbb{N}$ with $x_n \in U_{m_n}$. By passing to a subsequence if necessary, we can assume that $\{m_n\}_{n \in \mathbb{N}}$ is increasing. We thus have by definition of the projection

$$0 \leq \|x - P_{m_n} x\|_X = \inf_{u \in U_{mn}} \|x - u\|_X \leq \|x - x_n\|_X \to 0 \quad \text{as } n \to \infty.$$

Since the U_m are nested, $\{\|x - P_m x\|_X\}_{m \in \mathbb{N}}$ is decreasing; hence the full sequence must converge to 0. Lemma 15.12 then implies that

$$0 \leq \|x - \sum_{k=1}^m (x, e_k)_X e_k\|_X = \|x - P_m x\|_X \to 0 \quad \text{as } m \to \infty.$$

(ii) $\Rightarrow$ (iii): Let $x \in X$ be given. By (15.8), the partial sums $s_n := \sum_{k=1}^n (x, e_k)_X e_k$ then satisfy

$$\|s_n\|_X^2 = (s_n, s_n)_X = \sum_{k=1}^n |(x, e_k)_X|^2.$$

Now (ii) implies that $s_n \to x$ and hence that $\|s_n\|_X \to \|x\|_X$. Passing to the limit $n \to \infty$ on both sides thus yields (iii).

(iii) $\Rightarrow$ (ii): A similar calculation shows that for all $x \in X$, the partial sums s_n satisfy

$$(x, s_n)_X = \sum_{k=1}^n |(x, e_k)_X|^2 = \|s_n\|_X^2.$$

This implies

$$\|x - s_n\|_X^2 = \|x\|_X^2 - 2\operatorname{Re}(x, s_n)_X + \|s_n\|_X^2 = \|x\|_X^2 - \|s_n\|_X^2 \to 0$$

and hence (ii).

(ii) $\Rightarrow$ (i): Assume to the contrary that $\operatorname{lin} S$ is not dense in X. Then there exist an $x \in X$ and an $\varepsilon > 0$ such that $\|x - x_n\|_X > \varepsilon$ for all sequences $\{x_n\}_{n \in \mathbb{N}} \subset \operatorname{lin} S$. In particular, this holds for the sequence $\{s_n\}_{n \in \mathbb{N}}$ of partial sums, and hence (ii) is violated. $\square$

Note that (i) already implies that X is separable.

For example, the set $\{e_n : n \in \mathbb{N}\}$ of unit vectors in $\ell^2(\mathbb{F})$ is an orthonormal basis. It is slightly more complicated to verify that on $L^2((-\pi, \pi))$ for $\mathbb{F} = \mathbb{C}$, the functions

$$e_k(t) = \frac{1}{\sqrt{2\pi}} e^{ikt}, \quad k \in \mathbb{Z},$$

form an orthonormal basis. This allows writing any function $f \in L^2((-\pi, \pi))$ as

$$f(t) = \sum_{k \in \mathbb{Z}} c_k e_k(t), \quad c_k := (f, e_k)_{L^2} = \frac{1}{\sqrt{2\pi}} \int_{-\pi}^{\pi} f(t) e^{-ikt} \, dt.$$

This series is called the *Fourier series* of f. (In analogy, one also refers to the series in Theorem 15.14 (ii) as a *(generalized) Fourier series*, and correspondingly to $(x, e_k)_X$ as a *(generalized) Fourier coefficient*.)

In general, we have the following result.

Theorem 15.15

Let X be an infinite-dimensional Hilbert space. Then the following properties are equivalent:

(i) X is separable;

(ii) X contains a countable orthonormal basis.

Proof. (i) $\Rightarrow$ (ii): Let $\{x_n : n \in \mathbb{N}\}$ be dense in X. We now define inductively

$$\tilde{e}_n := x_n - \sum_{k=1}^{n-1} (x_n, e_k)_X \, e_k,$$

$$e_n := \begin{cases} \frac{\tilde{e}_n}{\|\tilde{e}_n\|_X} & \text{if } \tilde{e}_n \neq 0, \\ 0 & \text{if } \tilde{e}_n = 0. \end{cases}$$

Then $\|e_n\|_X = 1$ for all $n \in \mathbb{N}$ and $(e_n, e_k)_X = 0$ for all $k < n \in \mathbb{N}$, i.e., $\{e_n : n \in \mathbb{N}\}$ is an orthonormal system. Furthermore, $\mathrm{lin}\{e_n : n \in \mathbb{N}\} = \mathrm{lin}\{x_n : n \in \mathbb{N}\}$ is dense in X, and therefore $\{e_n : n \in \mathbb{N}\}$ is even an orthonormal basis.

(ii) $\Rightarrow$ (i): If $\{e_n : n \in \mathbb{N}\}$ is a countable orthonormal basis, then the set of all (finite) rational linear combinations is also countable as well as dense in $\mathrm{lin}\{e_n : n \in \mathbb{N}\}$ and therefore also in X. $\square$

The construction in the first step corresponds exactly to the *Gram–Schmidt process* from linear algebra.[3]

Theorem 15.15 implies the following remarkable result.

Corollary 15.16 (Fischer–Riesz theorem)

Let X be a separable infinite-dimensional Hilbert space over $\mathbb{F}$. Then X is isometrically isomorphic to $\ell^2(\mathbb{F})$.

---

[3]If X is not separable, the existence of an (in this case uncountable) orthonormal basis can still be shown using Zorn's lemma. For this it is necessary to use the fact that an orthonormal basis is a maximal orthonormal system, i.e., is not contained in a larger orthonormal system; see, e.g., [21, Theorem 4.22]. It is also possible to extend Theorem 15.14 to cover this situation, since even for an uncountable orthonormal system, at most countably many inner products $(x, e)_X$ are nonzero; see, e.g., [7, Theorem 4.13].

Proof. Theorem 15.15 guarantees that X contains a countable orthonormal basis $\{e_n : n \in \mathbb{N}\}$. We now use this to construct an isometric isomorphism $T : X \to \ell^2(\mathbb{F})$ by assigning to every $x \in X$ a sequence $Tx := \{y_k\}_{k \in \mathbb{N}}$ via $y_k := (x, e_k)_X$. Parseval's identity then implies that $Tx = y \in \ell^2(\mathbb{F})$ and that $\|Tx\|_{\ell^2} = \|x\|_X$. In particular, T is continuous. Furthermore, T is clearly linear and injective; it remains to show that it is surjective. To this end, let $y \in \ell^2(\mathbb{F})$ be given. The definition of the induced norm implies that $\{\sum_{k=1}^{n} |y_k|^2\}_{n \in \mathbb{N}}$ is a Cauchy sequence (in $\mathbb{R}$). It follows that $\{\sum_{k=1}^{n} y_k e_k\}_{n \in \mathbb{N}}$ is also a Cauchy sequence (in X), since for all $m < n \in \mathbb{N}$,

$$\left\| \sum_{k=m+1}^{n} y_k e_k \right\|_X^2 = \left(\sum_{k=m+1}^{n} y_k e_k, \sum_{j=m+1}^{n} y_j e_j \right)_X = \sum_{k=m+1}^{n} |y_k|^2.$$

Since X is complete, the series $\sum_{k=1}^{\infty} y_k e_k$ converges to some $x \in X$. The continuity of the inner product then yields

$$[Tx]_k = (x, e_k)_X = \lim_{n \to \infty} \left(\sum_{j=1}^{n} y_j e_j, e_k \right)_X = y_k \qquad \text{for all } k \in \mathbb{N}$$

and hence $Tx = y$. □

In a nutshell, this shows that *all* separable infinite-dimensional Hilbert spaces are isometrically isomorphic to each other.

Problems

Problem 15.1 *(Examples of Hilbert spaces)*

(i) Let $\mathbb{R}^{n \times n}$ be the space of real $n \times n$ matrices and set

$$(A, B) := \operatorname{tr}(AB^T) \qquad \text{for all } A, B \in \mathbb{R}^{n \times n},$$

where tr M is the trace and M^T the transposed matrix of $M \in \mathbb{R}^{n \times n}$. Show that this defines a Hilbert space. Deduce from this that

$$|\operatorname{tr}(AB^T)|^2 \leq \operatorname{tr}(AA^T) \operatorname{tr}(BB^T) \qquad \text{for all } A, B \in \mathbb{R}^{n \times n}.$$

(ii) Show that $(\ell^p(\mathbb{F}), \|\cdot\|_p)$ is a Hilbert space if and only if $p = 2$.
(iii) Show that $(C([a, b]), \|\cdot\|_\infty)$ is not a Hilbert space.

Problem 15.2 *(Orthogonality via norm)*
Let X be a Hilbert space (not necessarily over $\mathbb{R}$) and $x, y \in X$. Show that $(x, y)_X = 0$ if and only if

$$\|x + \alpha y\|_X = \|x - \alpha y\|_X \qquad \text{for all } \alpha \in \mathbb{F}.$$

Problem 15.3 *(Convergence and angles)*
Let X be a Hilbert space and $\{x_n\}_{n \in \mathbb{N}}, \{y_n\}_{n \in \mathbb{N}} \subset B_X$. Show that if $(x_n, y_n)_X \to 1$, then $\|x_n - y_n\|_X \to 0$.

Problem 15.4 *(Projections are nonexpansive)*
Let X be a Hilbert space and let $K \subset X$ be nonempty, convex, and closed. Show that

$$\|P_K(x) - P_K(\tilde{x})\|_X \le \|x - \tilde{x}\|_X \qquad \text{for all } x, \tilde{x} \in X.$$

Problem 15.5 *(Hahn–Banach in Hilbert spaces)*
Let X be a Hilbert space and U a closed subspace of X. Show that every continuous linear functional on U can be extended isometrically to X by explicitly constructing such an extension.
Hint: Use the projection P_U onto U.

The Riesz Representation Theorem

16

We now specialize the duality theory from Part III to Hilbert spaces. Recall that every Hilbert space X corresponds (via the induced norm) to a normed vector space, which in turn has a dual space X^*. We have already remarked on the formal similarity between the duality pairing $\langle \cdot, \cdot \rangle_X$ between X and X^* and the inner product $(\cdot, \cdot)_X$ on X; this similarity can be made rigorous.

Theorem 16.1 (Fréchet–Riesz representation theorem)
Let X be a Hilbert space. Then for every $x^* \in X^*$ there exists a unique Riesz representative *$x \in X$ such that*

$$\langle x^*, z \rangle_X = (z, x)_X \qquad \text{for all } z \in X.$$

Furthermore, $\|x\|_X = \|x^*\|_{X^*}$.

Proof. We first note that for fixed $x \in X$, the mapping $T_x : z \mapsto (z, x)_X$ is a continuous linear functional on X. We now show that

$$R_X : X \to X^*, \qquad x \mapsto T_x,$$

is a bijective isometric operator, i.e., that every $x^* \in X^*$ can be written as T_x for a unique $x \in X$. First, the definition of R_X and the Cauchy–Schwarz inequality yield

$$|\langle R_X(x), z \rangle_X| = |(z, x)_X| \le \|z\|_X \|x\|_X \qquad \text{for all } z \in X,$$

© Springer Nature Switzerland AG 2020
C. Clason, *Introduction to Functional Analysis*, Compact Textbooks
in Mathematics, https://doi.org/10.1007/978-3-030-52784-6_16

with equality for $z = x$. Taking the supremum over all $z \in B_X$ then shows that $\|R_X(x)\|_{X^*} = \|x\|_X$ for all $x \in X$. Hence R_X is isometric and thus injective.

To show surjectivity of R_X, let $x^* \in X^*$ be given. For $x^* = 0$ we simply choose $x = 0$. If $x^* \neq 0$, then $\ker x^*$ is a proper closed subspace of X by Theorem 8.3. This implies that $(\ker x^*)^\perp \neq \{0\}$ (otherwise we could use Corollary 15.10 to show that $\ker x^* = X$). We can thus find an $x \in (\ker x^*)^\perp \setminus \{0\}$. In particular, we then have $\langle x^*, x\rangle_X \neq 0$, since otherwise $x \in (\ker x^*)^\perp \cap \ker x^*$. Hence Theorem 15.9 (iv) yields

$$x = P_{\ker x^*}(x) = x - P_{(\ker x^*)^\perp}(x) = x - x = 0,$$

in contradiction to $x \neq 0$. Let now $z \in X$ be arbitrary. Since for all $\lambda \in \mathbb{F}$,

$$\langle x^*, z - \lambda x\rangle_X = \langle x^*, z\rangle_X - \lambda\langle x^*, x\rangle_X,$$

we can set $\lambda_z := \frac{\langle x^*, z\rangle_X}{\langle x^*, x\rangle_X}$ and obtain $z - \lambda_z x \in \ker x^*$. Since $x \in (\ker x^*)^\perp$, it follows that $(z - \lambda_z x, x)_X = 0$. Together with the definition of λ_z, this implies that

$$\frac{\langle x^*, z\rangle_X}{\langle x^*, x\rangle_X} = \lambda_z = \frac{(z, x)_X}{(x, x)_X}.$$

Rearranging then yields

$$\langle x^*, z\rangle_X = \left(z, \frac{\overline{\langle x^*, x\rangle_X}}{\|x\|_X^2} x\right)_X,$$

i.e., $x^* = R_X\left(\frac{\overline{\langle x^*, x\rangle_X}}{\|x\|_X^2} x\right)$. Hence R_X is surjective. $\square$

The mapping $R_X : X \to X^*$ is called the *Riesz isomorphism*, even though it is linear and hence an isomorphism only for $\mathbb{F} = \mathbb{R}$. If $\mathbb{F} = \mathbb{C}$, it is merely *antilinear*: the sesquilinearity of the inner product implies for all $x, y \in X, \alpha \in \mathbb{F}$, and arbitrary $z \in X$ that

$$\langle R_X(\alpha x + y), z\rangle_X = (z, \alpha x + y)_X = \overline{\alpha}\,(z, x)_X + (z, y)_X$$
$$= \langle \overline{\alpha} R_X(x) + R_X(y), z\rangle_X.$$

The Riesz isomorphism can be used to transfer properties between a Hilbert space and its dual space.

Corollary 16.2
Let X be a Hilbert space. Then
  (i) X^* is a Hilbert space;
  (ii) X is reflexive.

Proof.

(i) It is straightforward to use the (anti)linearity and bijectivity of R_X to verify that

$$\left(x^*, y^*\right)_{X^*} := \left(R_X^{-1} y^*, R_X^{-1} x^*\right)_X \qquad \text{for all } x^*, y^* \in X^*$$

defines an inner product on X^*, which is therefore a pre-Hilbert space. Furthermore, since R_X is an isometry,

$$\|x^*\|_{X^*}^2 = \|R_X^{-1} x^*\|_X^2 = \left(R_X^{-1} x^*, R_X^{-1} x^*\right)_X = \left(x^*, x^*\right)_{X^*} \qquad \text{for all } x^* \in X^*,$$

i.e., the operator norm is induced by this inner product. Since dual spaces are always complete with respect to this norm, X^* is a Hilbert space.

(ii) We have to show that the canonical embedding $J_X : X \to X^{**}$ is surjective, which we do by showing that $J_X = R_{X^*} \circ R_X$. Let $x \in X$ and $x^* \in X^*$ be arbitrary. Then the definitions of the Riesz isomorphism, of the inner product in X^*, and of the canonical embedding yield

$$\langle R_{X^*} R_X x, x^* \rangle_{X^*} = \left(x^*, R_X x\right)_{X^*} = \left(x, R_X^{-1} x^*\right)_X = \langle x^*, x \rangle_X = \langle J_X x, x^* \rangle_{X^*}.$$

Hence J_X is a composition of bijective mappings and therefore surjective.   □

The Riesz isomorphism allows characterizing weak convergence via the inner product: the bijectivity of R_X directly implies that

$$x_n \rightharpoonup x \qquad \text{if and only if} \qquad (x_n, z)_X \to (x, z)_X \qquad \text{for all } z \in X.$$

Hence weak convergence in Hilbert spaces does not require dual spaces, and the gap to strong convergence can be closed partially.

Corollary 16.3
Let X be a Hilbert space and $\{x_n\}_{n\in\mathbb{N}} \subset X$. Then the following properties are equivalent:

(i) $x_n \to x$;

(ii) $x_n \rightharpoonup x$ and $\|x_n\|_X \to \|x\|_X$.

Proof. We already know that in normed vector spaces, strong convergence implies weak convergence and that the norm is continuous. Conversely, weak convergence and convergence of the induced norm in a Hilbert space imply that

$$\|x_n - x\|_X^2 = \|x_n\|_X^2 - 2\,\mathrm{Re}\,(x_n, x)_X + \|x\|_X^2 \to \|x\|_X^2 - 2\,(x, x)_X + \|x\|_X^2 = 0. \qquad \square$$

Even in Banach spaces, convergence according to (ii) can be useful as a (in this case distinct) concept; it is then referred to as *strict convergence*.

Similarly, we can use the Riesz isomorphism to "pull back" the adjoint operator into X: for Hilbert spaces X and Y and for a linear operator $T \in L(X, Y)$, we define the *Hilbert-space adjoint operator*

$$T^\star : Y \to X, \qquad T^\star y = R_X^{-1} T^* R_Y y,$$

where $T^* : Y^* \to X^*$ is the already introduced (Banach-space) adjoint operator. (Despite the similar notation, there will be no danger of confusion in the following.) It follows from the definition of the Riesz isomorphism and of the adjoint operator that for all $x \in X$ and $y \in Y$,

$$(Tx, y)_Y = \langle R_Y y, Tx \rangle_Y = \langle T^* R_Y y, x \rangle_X = \left(x, R_X^{-1} T^* R_Y \right)_X$$
$$= \left(x, T^\star y \right)_X. \tag{16.1}$$

The following calculus is a direct consequence of the definition as well.

Lemma 16.4
Let X, Y, and Z be Hilbert spaces, $S, T \in L(X, Y)$, and $R \in L(Y, Z)$. Then

(i) $(S + T)^\star = S^\star + T^\star$;

(ii) $(\lambda T)^\star = \overline{\lambda} T^\star$ for all $\lambda \in \mathbb{F}$;

(iii) $(R \circ T)^\star = T^\star \circ R^\star$.

Exactly as for Theorem 9.6 (simply using Corollary 15.10 instead of Corollary 8.8), we obtain the following results on the orthogonal complements of null spaces and ranges of Hilbert-space adjoints. From now on, for brevity we will write $T^\star T := T^\star \circ T$.

Lemma 16.5

Let X and Y be Hilbert spaces and $T \in L(X, Y)$. Then

   (i) $T^{\star\star} = T$;

   (ii) $\|T^\star\|_{L(Y,X)} = \|T\|_{L(X,Y)}$;

   (iii) $\|T^\star T\|_{L(X,X)} = \|T\|^2_{L(X,Y)}$.

Proof.

 (i) We immediately obtain from (16.1) that for all $x \in X$ and $y \in Y$,

$$\left(y, T^{\star\star}x\right)_Y = \left(T^\star y, x\right)_X = \overline{(x, T^\star y)_X} = \overline{(Tx, y)_Y} = (y, Tx)_Y .$$

 (ii) First, we have for all $y \in Y$ that

$$\|T^\star y\|_X = \|R_X^{-1} T^* R_Y y\|_X = \|T^* R_Y y\|_{X^*} \leq \|T^*\|_{L(Y^*, X^*)} \|R_Y y\|_{Y^*} = \|T\|_{L(X,Y)} \|y\|_Y$$

since Riesz isomorphisms and the mapping $T \mapsto T^*$ are isometries (by Theorem 16.1 and Lemma 9.1, respectively). Taking the supremum over all $y \in B_Y$ then yields $\|T^\star\|_{L(Y,X)} \leq \|T\|_{L(X,Y)}$. Furthermore, (i) implies that $\|T\|_{L(X,Y)} = \|T^{\star\star}\|_{L(X,Y)} \leq \|T^\star\|_{L(Y,X)}$ and hence the claim.

(iii) Corollary 4.6 together with (ii) yields

$$\|T^\star T\|_{L(X,X)} \leq \|T^\star\|_{L(Y,X)} \|T\|_{L(X,Y)} = \|T\|^2_{L(X,Y)}.$$

The reverse inequality follows from

$$\|Tx\|^2_Y = (Tx, Tx)_Y = \left(x, T^\star Tx\right)_X \leq \|T^\star T\|_{L(X,X)} \|x\|^2_X$$

and taking the supremum over all $x \in B_X$.   □

The Riesz isomorphism thus allows building a complete duality theory using only elements of X. It is therefore common not to distinguish between elements $x^* \in X^*$ and their Riesz representatives $R_X^{-1} x^* \in X$, i.e., to treat R_X as the identity; in other words, one *identifies* X^* with X. In particular, one usually does not distinguish between Banach-space and Hilbert-space adjoints. However, this is not always reasonable. One prominent example involves two Hilbert spaces X and

Y, endowed with different inner products, where X is continuously embedded and dense in Y. In this case, Y^* is continuously embedded into X^*; identifying Y^* with Y (i.e., taking R_Y as the identity) creates a *Gelfand triple* $X \hookrightarrow Y \cong Y^* \hookrightarrow X^*$. If we now also identify X^* with X, these embeddings become meaningless; we thus have to make a decision which dual space to identify with its primal. (Of course, there still is a Riesz isomorphism R_X; it just cannot be treated as the identity.) This situation is especially relevant in the theory of partial differential equations (where R_X^{-1} itself often corresponds to the solution of a differential equation.)

Problems

Problem 16.1 *(Lax–Milgram theorem revisited)*
Let X be a Hilbert space and $a : X \times X \to \mathbb{F}$ a sesquilinear form. Show that if there exist constants $C, \gamma > 0$ such that

$$|a(x, y)| \leq C \|x\|_X \|y\|_X \quad \text{for all } x, y \in X,$$
$$\operatorname{Re} a(x, x) \geq \gamma \|x\|_X^2 \qquad \text{for all } x \in X,$$

then there exists a unique linear operator $A : X \to X$ such that

$$a(x, y) = (Ax, y)_X \qquad \text{for all } x, y \in X.$$

Deduce from this that A is invertible with $\|A^{-1}\|_{L(X)} \leq \gamma^{-1}$.

Problem 16.2 *(Series in Hilbert spaces)*
Let X be a Hilbert space and $\{x_n\}_{n \in \mathbb{N}} \subset X$ an orthogonal system, i.e., $(x_i, x_j)_X = 0$ for all $i \neq j$. Show that the following properties are equivalent:

(i) $\sum_{n=1}^{\infty} x_n$ converges;
(ii) $\sum_{n=1}^{\infty} \|x_n\|_X^2$ converges;
(iii) $\sum_{n=1}^{\infty} x_n$ converges weakly.

(As in normed vector spaces, a series is called (weakly) convergent in a Hilbert space X if the corresponding sequence of partial sums converges (weakly) in X.)

Problem 16.3 *(Weak convergence of orthonormal systems)*
Let X be a Hilbert space and $\{e_n : n \in \mathbb{N}\}$ an orthonormal system. Show that $e_n \rightharpoonup 0$.

Problem 16.4 *(Hilbert-space adjoint projections)*
Let X be a Hilbert space, $U \subset X$ a closed subspace, and $P_U : X \to U$ the metric projection onto U.

(i) Determine its Hilbert-space adjoint $P^\star : U \to X$.
(ii) Deduce that the extension of a continuous linear functional from U to X constructed in Problem 15.5 is unique.

Problem 16.5 *(A Gelfand triple)*

Let $H = \ell^2(\mathbb{F})$ and

$$V = \left\{ x \in \ell^\infty(\mathbb{F}) : \sum_{k=1}^\infty k^2 x_k^2 < \infty \right\}, \qquad (x, y)_V := \sum_{k=1}^\infty k^2 x_k y_k \qquad \text{for } x, y \in V.$$

Show that $V \hookrightarrow H \hookrightarrow V^\star$, and determine the Riesz isomorphism R_V.

Spectral Decomposition in Hilbert Spaces $\quad$ **17**

One of the central results in linear algebra is the existence of the *spectral decomposition*: every normal matrix is diagonalizable, i.e., can be represented with respect to a basis of eigenvectors as a diagonal matrix. An analogous result is possible for compact operators on Hilbert spaces, closing (for such operators) the gap between linear algebra and functional analysis.

Throughout this chapter, let X be a Hilbert space. We call $T \in L(X) := L(X, X)$ *normal* if $T^\star T = TT^\star$, and *self-adjoint* if $T = T^\star$. Clearly, every self-adjoint operator is normal, and $T^\star T$ and $TT^\star$ are always self-adjoint for any $T \in L(X)$.

We now study the eigenvalues of normal and of self-adjoint operators.

Theorem 17.1

Let $T \in L(X)$ be normal. Then

(i) $\|Tx\|_X = \|T^\star x\|_X$ for all $x \in X$;
(ii) $\ker T = \ker T^\star$;
(iii) $Tx = \lambda x$ if and only if $T^\star x = \bar{\lambda} x$.

Proof.

(i) If T is normal, then we have for all $x \in X$ that

$$\|Tx\|_X^2 = (Tx, Tx)_X = \left(x, T^\star T x\right)_X = \left(x, T T^\star x\right)_X = \left(T^\star x, T^\star x\right)_X = \|T^\star x\|_X^2,$$

using $T = T^{\star\star}$ for the next-to-last equality.

© Springer Nature Switzerland AG 2020
C. Clason, *Introduction to Functional Analysis*, Compact Textbooks
in Mathematics, https://doi.org/10.1007/978-3-030-52784-6_17

(ii) It follows from (i) that $Tx = 0$ if and only if $T^\star x = 0$.

(iii) It is straightforward to verify that normality of T implies normality of $\lambda \operatorname{Id} - T$ for all $\lambda \in \mathbb{F}$. Hence (ii) together with Lemma 16.4 (ii) implies that

$$\ker(\lambda \operatorname{Id} - T) = \ker(\lambda \operatorname{Id} - T)^\star = \ker(\overline{\lambda} \operatorname{Id} - T^\star). \qquad \square$$

In other words, (iii) states that λ is an eigenvalue of T if and only if $\overline{\lambda}$ is an eigenvalue of $T^\star$.

Corollary 17.2

Let $T \in L(X)$ be normal. Then eigenvectors corresponding to different eigenvalues are orthogonal.

Proof. Let $\lambda_1, \lambda_2 \in \mathbb{F}$ and $x_1, x_2 \in X$ be given with $Tx_1 = \lambda_1 x_1$ and $Tx_2 = \lambda_2 x_2$ as well as $\lambda_1 \neq \lambda_2$. Then Theorem 17.1 (iii) together with the sesquilinearity of the inner product yields

$$\lambda_1 (x_1, x_2)_X = (\lambda_1 x_1, x_2)_X = (Tx_1, x_2)_X = \left(x_1, T^\star x_2\right)_X = \left(x_1, \overline{\lambda_2} x_2\right)_X = \lambda_2 (x_1, x_2)_X \,,$$

which is possible only for $(x_1, x_2)_X = 0$. $\qquad \square$

For normal operators, we also obtain a sharper estimate for the spectral radius $r(T) := \sup_{\lambda \in \sigma(T)} |\lambda|$; compare Theorem 14.6.

Theorem 17.3

Let $\mathbb{F} = \mathbb{C}$ and $T \in L(X)$ be normal. Then $r(T) = \|T\|_{L(X)}$.

Proof. We first show by induction that $\|T^n\|_{L(X)} = \|T\|_{L(X)}^n$ for all $n \in \mathbb{N}$. For $n = 1$ the claim is trivial; for $n = 2$, Lemma 16.4 (iii) implies that

$$\|T^2\|_{L(X)}^2 = \|(T^2)^\star (T)^2\|_{L(X)} = \|(T^\star T)^\star (T^\star T)\|_{L(X)} = \|T^\star T\|_{L(X)}^2$$

$$= \left(\|T\|_{L(X)}^2\right)^2, \tag{17.1}$$

where we have used the normality of T for the second equality. For the induction step $n \mapsto n + 1$ for $n \geq 2$, we first note that a straightforward calculation shows that T^n is normal as well. We can thus apply the induction assumption for $n \geq 2$ to T as well as to T^n to obtain

$$\|T\|_{L(X)}^{2n} = (\|T\|_{L(X)}^n)^2 = \|T^n\|_{L(X)}^2 = \|(T^n)^2\|_{L(X)} \leq \|T^{n+1}\|_{L(X)} \|T\|_{L(X)}^{n-1}.$$

Dividing by $\|T\|_{L(X)}^{n-1} \neq 0$ (otherwise the claim holds trivially) together with Corollary 4.6 then yields

$$\|T\|_{L(X)}^{n+1} \leq \|T^{n+1}\|_{L(X)} \leq \|T\|_{L(X)}^{n+1}$$

and hence $\|T\|_{L(X)}^{n+1} = \|T^{n+1}\|_{L(X)}$.

We thus obtain from Theorem 14.6 that

$$r(T) = \lim_{n \to \infty} \|T^n\|_{L(X)}^{1/n} = \|T\|_{L(X)}. \qquad \square$$

Since the spectrum is always compact by Theorem 14.4 and for compact operators consists only of eigenvalues by the little spectral theorem (Theorem 14.7), this implies the following useful result.

> **Corollary 17.4**
>
> Let $X \neq \{0\}$ be a Hilbert space over $\mathbb{F} = \mathbb{C}$ and let $T \in L(X)$ be compact and normal. Then there exists an eigenvalue $\lambda \in \sigma_p(T)$ with $|\lambda| = \|T\|_{L(X)}$.

We can obtain the same result for $\mathbb{F} = \mathbb{R}$ if we consider self-adjoint operators. Instead of Theorem 14.6, we will then use the following characterization of the operator norm.

> **Theorem 17.5**
>
> Let $T \in L(X)$ be self-adjoint. Then $\|T\|_{L(X)} = \sup_{x \in B_X} |(Tx, x)_X|$.

Proof. The Cauchy–Schwarz inequality together with Lemma 4.3 (i) immediately yields

$$M := \sup_{x \in B_X} |(Tx, x)_X| \leq \sup_{x \in B_X} \|Tx\|_X = \|T\|_{L(X)}.$$

For the reverse inequality, we write $\|Tx\|_X^2 = (Tx, Tx)_X = \left(\alpha Tx, \alpha^{-1} Tx\right)_X$ for arbitrary $\alpha > 0$. We now add and subtract terms and use the self-adjointness of T in order to estimate

$$
\begin{aligned}
4\|Tx\|_X^2 &= \left(T(\alpha x + \alpha^{-1} Tx), \alpha x + \alpha^{-1} Tx\right)_X - \left(T(\alpha x - \alpha^{-1} Tx), \alpha x - \alpha^{-1} Tx\right)_X \\
&= \left(T\left(\frac{\alpha x + \alpha^{-1} Tx}{\|\alpha x + \alpha^{-1} Tx\|_X}\right), \frac{\alpha x + \alpha^{-1} Tx}{\|\alpha x + \alpha^{-1} Tx\|_X}\right)_X \|\alpha x + \alpha^{-1} Tx\|_X^2 \\
&\quad - \left(T\left(\frac{\alpha x - \alpha^{-1} Tx}{\|\alpha x - \alpha^{-1} Tx\|_X}\right), \frac{\alpha x - \alpha^{-1} Tx}{\|\alpha x - \alpha^{-1} Tx\|_X}\right)_X \|\alpha x - \alpha^{-1} Tx\|_X^2 \\
&\leq M\left(\|\alpha x + \alpha^{-1} Tx\|_X^2 + \|\alpha x - \alpha^{-1} Tx\|_X^2\right) \\
&= 2M\left(\alpha^2 \|x\|_X^2 + \alpha^{-2}\|Tx\|_X^2\right),
\end{aligned}
$$

where we have used the parallelogram identity (Theorem 15.4) in the last step. If $Tx \neq 0$, we can choose $\alpha^2 := \frac{\|Tx\|_X}{\|x\|_X} > 0$ and divide by $\|Tx\|_X > 0$ to obtain

$$
\|Tx\|_X \leq M\|x\|_X \qquad \text{for all } x \in X
$$

(which holds trivially for $Tx = 0$). Taking the supremum over all $x \in B_X$ then yields

$$
\|T\|_{L(X)} = \sup_{x \in B_X} \|Tx\|_X \leq M. \qquad \qquad \square
$$

We can now show the counterpart of Corollary 17.4 for self-adjoint real operators.

Corollary 17.6

Let $X \neq \{0\}$ be a Hilbert space over $\mathbb{F} = \mathbb{R}$ and let $T \in L(X)$ be compact and self-adjoint. Then there exists an eigenvalue $\lambda \in \sigma_p(T)$ with $|\lambda| = \|T\|_{L(X)}$.

Proof. First, it follows from Theorem 17.5 that there exists a sequence $\{x_n\}_{n \in \mathbb{N}} \subset B_X$ such that $|(Tx_n, x_n)_X| \to \|T\|_{L(X)}$. By passing to a subsequence (again denoted by $\{x_n\}_{n \in \mathbb{N}}$) if necessary, we can assume that $(Tx_n, x_n)_X \to \lambda$ with $|\lambda| = \|T\|_{L(X)}$. This implies that

$$
\begin{aligned}
\|Tx_n - \lambda x_n\|_X^2 &= (Tx_n - \lambda x_n, Tx_n - \lambda x_n)_X \\
&= \|Tx_n\|_X^2 - 2\lambda(Tx_n, x_n)_X + \lambda^2 \|x_n\|_X^2 \\
&\leq \|T\|_{L(X)}^2 - 2\lambda(Tx_n, x_n)_X + \lambda^2 \to 0. \qquad (17.2)
\end{aligned}
$$

Since B_X is bounded and T is compact, we can find a convergent subsequence (again denoted by $\{x_n\}_{n \in \mathbb{N}}$) such that

$$T x_n \to y, \qquad (T x_n, x_n)_X \to \lambda.$$

Together with (17.2) and the continuity of T, we obtain from this that

$$y = \lim_{n \to \infty} T x_n = \lim_{n \to \infty} \lambda x_n \qquad \text{and} \qquad T y = \lambda (\lim_{n \to \infty} T x_n) = \lambda y.$$

If $y \neq 0$, then λ is the desired eigenvalue of T. Otherwise, $\{T x_n\}_{n \in \mathbb{N}}$ is a null sequence, and Theorem 17.5 implies that $\|T\|_{L(X)} = \lim_{n \to \infty} |(T x_n, x_n)_X| = 0$ and hence that $T = 0$. But in this case, the claim holds trivially.   □

To top it all off, we can now show that every compact normal or self-adjoint operator on a Hilbert space admits a spectral decomposition.

Theorem 17.7 (Spectral theorem)
Let X be a Hilbert space over $\mathbb{F}$ and let $T \in L(X)$ be compact and normal (if $\mathbb{F} = \mathbb{C}$) or self-adjoint (if $\mathbb{F} = \mathbb{R}$). Then there exists a (possibly finite) orthonormal system $\{e_n : n \in \mathbb{N}\} \subset X$ of eigenvectors of T and (in this case also finite) null sequence $\{\lambda_n\}_{n \in \mathbb{N}} \subset \mathbb{F}$ of corresponding eigenvalues satisfying the spectral decomposition

$$T x = \sum_{n=1}^{\infty} \lambda_n (x, e_n)_X \, e_n \qquad \text{for all } x \in X. \tag{17.3}$$

Furthermore, $\{e_n : n \in \mathbb{N}\}$ is an orthonormal basis of $(\ker T)^{\perp}$.

Proof. We proceed by induction. Set $X_1 := X$ and $T_1 := T$. If $X_1 = \{0\}$ or $T_1 = 0$, the claim holds with $N = 0$. (In this case we set $\lambda_1 = 0$.) Otherwise, Corollary 17.4 (if $\mathbb{F} = \mathbb{C}$) or Corollary 17.6 (if $\mathbb{F} = \mathbb{R}$) guarantees the existence of an eigenvalue $\lambda_1 \in \sigma_p(T_1)$ with $|\lambda_1| = \|T_1\|_{L(X)}$. Let $e_1 \in X_1$ be a corresponding eigenvector with $\|e_1\|_X = 1$ and set $X_2 := (\mathrm{lin}\{e_1\})^{\perp} \subset X_1$. By Theorem 17.1 (iii), we then have for all $x \in X_2$ that

$$(T_1 x, e_1)_X = (x, T_1^{\star} e_1)_X = (x, \overline{\lambda_1} e_1)_X = \lambda_1 (x, e_1)_X = 0$$

and hence that $T_1 x \in X_2$. We now set $T_2 := T_1|_{X_2} \in L(X_2)$, which as the restriction of a compact normal (or self-adjoint) operator is compact and normal (or self-adjoint) as well. Furthermore,

$$\|T_2\|_{L(X_2)} = \sup_{x \in B_{X_2}} \|T_2 x\|_X = \sup_{x \in B_{X_2}} \|T_1 x\|_X \leq \sup_{x \in B_{X_1}} \|T_1 x\|_X = \|T_1\|_{L(X_1)}.$$

We now repeat this construction with $X_n = (\lin\{e_1, \ldots, e_{n-1}\})^{\perp}$ for $n \geq 0$ to obtain an orthonormal system $\{e_n : n \in \mathbb{N}\}$ and a sequence $\{\lambda_n\}_{n \in \mathbb{N}}$ with $T e_n = \lambda_n e_n$ and $|\lambda_n| = \|T_n\|_{L(X_n)}$. In particular, $\{|\lambda_n|\}_{n \in \mathbb{N}}$ is decreasing. If $X_m = \{0\}$ or $T_m = 0$ for some $m \in \mathbb{N}$, these sequences terminate with $N = m - 1$ and $\lambda_n = 0$ for $n \geq N + 1$. Otherwise, $\{T e_n\}_{n \in \mathbb{N}}$ contains a convergent subsequence since $\{e_n : n \in \mathbb{N}\} \subset B_X$ is bounded and T is compact. Furthermore, $\{e_n : n \in \mathbb{N}\}$ is an orthonormal system and hence

$$|\lambda_{n_k}|^2 \leq |\lambda_{n_k}|^2 + |\lambda_{n_l}|^2 = \|\lambda_{n_k} e_{n_k} - \lambda_{n_l} e_{n_l}\|_X^2 = \|T e_{n_k} - T e_{n_l}\|_X^2 \to 0$$

as $k, l \to \infty$ along this subsequence. Since $\{|\lambda_n|\}_{n \in \mathbb{N}}$ is decreasing, we even have $|\lambda_n| \to 0$ along the full sequence.

Consider now for $x \in X$ the orthogonal projection onto X_n, i.e.,

$$x_n := P_{X_n}(x) = x - P_{\lin\{e_1, \ldots, e_{n-1}\}}(x) = x - \sum_{k=1}^{n-1} (x, e_k)_X \, e_k.$$

Then

$$\|Tx - \sum_{k=1}^{n-1} \lambda_k (x, e_k)_X \, e_k\|_X = \|Tx - \sum_{k=1}^{n-1} (x, e_k)_X \, T e_k\|_X$$

$$= \|T x_n\|_X = \|T_n x_n\|_X$$

$$\leq \|T_n\|_{L(X_n)} \|x_n\|_X \leq |\lambda_n| \|x\|_X \to 0$$

since $|\lambda_n| = \|T_n\|_{L(X_n)}$ by construction and $\|P_{x_n}(x)\|_X \leq \|x\|_X$ for all $x \in X$. This yields the spectral decomposition (17.3).

It remains to show that $S := \lin\{e_n : n \in \mathbb{N}\}$ is dense in $(\ker T)^{\perp}$, from which the final claim follows by Theorem 15.14. To this end, let $x \in S^{\perp}$. Then by construction, $x \in X_n$ for all $n \in \mathbb{N}$ (or for all $n \leq N + 1$ if the above process terminated), and hence

$$\|Tx\|_X = \|T_n x\|_X \leq \|T_n\|_{L(X_n)} \|x\|_X = |\lambda_n| \|x\|_X \to 0$$

as $n \to \infty$ (or for $n = N + 1$). This yields $Tx = 0$ and thus $x \in \ker T$. Conversely, $x \in \ker T$ implies that

$$0 = (Tx, e_n)_X = (x, T^{\star} e_n)_X = \lambda_n (x, e_n)_X \quad \text{for all } n \in \mathbb{N}$$

and thus $x \in S^{\perp}$. Hence $S^{\perp} = \ker T$, and Corollary 15.10 yields $\cl S = (S^{\perp})^{\perp} = (\ker T)^{\perp}$ and therefore the claim. $\square$

Problems

Problem 17.1 *(Characterization of normal operators)*
Show that $T \in L(X)$ is normal if *and only if* $\|Tx\|_X = \|T^\star x\|_X$.

Problem 17.2 *(Characterization of complex self-adjoint operators)*
Let X be a Hilbert space over $\mathbb{C}$ and $T \in L(X)$. Show that T is self-adjoint if and only if

$$(Tx, x)_X \in \mathbb{R} \qquad \text{for all } x \in X.$$

Hint: For the converse implication, consider $x + \lambda y$ for some $\lambda \in \mathbb{C}$ (this is called a polarization argument).

Problem 17.3 *(Decomposition of complex operators)*
Let X be a Hilbert space over $\mathbb{C}$ and $T \in L(X)$. Show that there exists a unique pair of self-adjoint operators $T_1, T_2 \in L(X)$ such that

$$T = T_1 + iT_2.$$

Hint: Construct T_1, T_2 explicitly using T and $T^\star$.

Problem 17.4 *(Hellinger–Toeplitz theorem)*
Let X be a Hilbert space and let $T : X \to X$ be a self-adjoint linear operator. Show that T is continuous.

Problem 17.5 *(Reverse spectral theorem)*
Let X be a Hilbert space, $\{e_n : n \in \mathbb{N}\}$ an orthonormal basis, and $\{\lambda_n\}_{n\in\mathbb{N}} \in \ell^\infty(\mathbb{R})$. Let

$$T : X \to X, \qquad x \mapsto \sum_{n\in\mathbb{N}} \lambda_n (x, e_n)_X e_n.$$

Show that

(i) T is linear and continuous;
(ii) T is normal;
(iii) T is compact if and only if $\lambda_n \to 0$.

Determine the eigenvalues and eigenvectors of T.

References

1. Alt, H.W.: Linear Functional Analysis. Universitext. Springer, Berlin (2016). https://doi.org/10.1007/978-1-4471-7280-2
2. Bourbaki, N.: Topological Vector Spaces. Chapters 1–5. Elements of Mathematics (Berlin). Springer, Berlin (1987). https://doi.org/10.1007/978-3-642-61715-7. Translated from the French by H.G. Eggleston and S. Madan
3. Brezis, H.: Functional Analysis, Sobolev Spaces and Partial Differential Equations. Springer, Berlin (2011). https://doi.org/10.1007/978-0-387-70914-7
4. Brokate, M.: Funktionalanalysis (2013). http://www-m6.ma.tum.de/~brokate/fun_ws13.pdf. Lecture notes, Zentrum Mathematik, TU München
5. Brokate, M., Kersting, G.: Measure and Integral. Compact Textbooks in Mathematics. Birkhäuser/Springer, Cham (2015). https://doi.org/10.1007/978-3-319-15365-0. Revised and updated translation of the 2011 German original
6. Clason, C.: Regularization of Inverse Problems (2020). https://arxiv.org/abs/2001.00617. Lecture notes
7. Conway, J.B.: A Course in Functional Analysis. Graduate Texts in Mathematics, vol. 96, 2nd edn. Springer, New York (1990). https://doi.org/10.1007/978-1-4757-4383-8
8. Conway, J.B.: A Course in Point Set Topology. Undergraduate Texts in Mathematics. Springer, Cham (2014). https://doi.org/10.1007/978-3-319-02368-7
9. Dieudonné, J.: History of Functional Analysis. North-Holland Mathematics Studies, vol. 49. North-Holland Publishing, Amsterdam-New York (1981). Notas de Matemática [Mathematical Notes], 77
10. Enflo, P.: A counterexample to the approximation problem in Banach spaces. Acta Math. **130**, 309–317 (1973). https://doi.org/10.1007/BF02392270
11. Engl, H.W., Hanke, M., Neubauer, A.: Regularization of Inverse Problems. Mathematics and Its Applications, vol. 375. Kluwer Academic Publishers Group, Dordrecht (1996). https://doi.org/10.1007/978-94-009-1740-8
12. Heuser, H.: Funktionalanalysis, 4th edn. Vieweg+Teubner, Wiesbaden (2006). https://doi.org/10.1007/978-3-322-96755-8
13. Jordan, P., von Neumann, J.: On inner products in linear, metric spaces. Ann. Math. (2) **36**(3), 719–723 (1935). https://doi.org/10.2307/1968653
14. Kaballo, W.: Grundkurs Funktionalanalysis, 2nd edn. Springer Spektrum, Heidelberg (2018). https://doi.org/10.1007/978-3-662-54748-9
15. Kreyszig, E.: Introductory Functional Analysis with Applications. Wiley Classics Library. Wiley, New York (1989)

© Springer Nature Switzerland AG 2020
C. Clason, *Introduction to Functional Analysis*, Compact Textbooks in Mathematics, https://doi.org/10.1007/978-3-030-52784-6

16. Lang, S.: Complex Analysis. Graduate Texts in Mathematics, vol. 103, 4th edn. Springer, New York (1999). https://doi.org/10.1007/978-1-4757-3083-8
17. Luxemburg, W.A.J., Väth, M.: The existence of non-trivial bounded functionals implies the Hahn-Banach extension theorem. Z. Anal. Anwendungen **20**(2), 267–279 (2001). https://doi.org/10.4171/zaa/1015
18. Pietsch, A.: History of Banach Spaces and Linear Operators. Birkhäuser, Boston (2007). https://doi.org/10.1007/978-0-8176-4596-0
19. Raman-Sundström, M.: A pedagogical history of compactness. Am. Math. Mon. **122**(7), 619–635 (2015). https://doi.org/10.4169/amer.math.monthly.122.7.619
20. Rudin, W.: Principles of Mathematical Analysis, 3rd edn. McGraw-Hill, New York (1976). International Series in Pure and Applied Mathematics
21. Rudin, W.: Real and Complex Analysis, 3rd edn. McGraw-Hill, New York (1987). International Series in Pure and Applied Mathematics
22. Rudin, W.: Functional Analysis, 2nd edn. McGraw-Hill, New York (1991)
23. Rynne, B.P., Youngson, M.A.: Linear Functional Analysis, 2nd edn. Springer Undergraduate Mathematics Series. Springer, London (2008). https://doi.org/10.1007/978-1-84800-005-6
24. Schechter, E.: Handbook of Analysis and Its Foundations. Academic Press, San Diego (1997)
25. Steen, L.A., Seebach Jr., J.A.: Counterexamples in Topology. Dover Publications, Mineola (1995). Reprint of the second (1978) edition
26. Voigt, J.: A Course on Topological Vector Spaces. Compact Textbooks in Mathematics. Birkhäuser/Springer, Cham (2020). https://doi.org/10.1007/978-3-030-32945-7
27. Wachsmuth, G.: Funktionalanalysis (2013). Lecture notes, Fakultät für Mathematik, TU Chemnitz
28. Werner, D.: Funktionalanalysis, 8th edn. Springer Spektrum, Berlin (2018). https://doi.org/10.1007/978-3-662-55407-4
29. Yosida, K.: Functional Analysis. Classics in Mathematics. Springer, Berlin (1995). https://doi.org/10.1007/978-3-642-61859-8. Reprint of the sixth (1980) edition
30. Zeidler, E.: Applied Functional Analysis: Applications to Mathematical Physics. Applied Mathematical Sciences, vol. 108. Springer, New York (1995). https://doi.org/10.1007/978-1-4612-0821-1
31. Zeidler, E.: Applied Functional Analysis: Main Principles and Their Applications. Applied Mathematical Sciences, vol. 109. Springer, New York (1995). https://doi.org/10.1007/978-1-4612-0821-1

Index

Page numbers with 'N' refer to footnotes, numbers with 'P' to problems.

A
Annihilator, 67
Argument
    density, 40
    polarization, 163P
    subsequence–subsequence, 8N

B
Ball
    closed, 4
    open, 4
    unit, 22
    p-, 32P
Basis, 23
    orthonormal, 144

C
Closure, 5
Codimension, 120
Coefficient, 23
    Fourier, 146
Complement, orthogonal, 138
Convergence, 5
    strict, 152
    strong, 100
    weak, 99
    weak-∗, 99
Core, 46

D
Decomposition, spectral, 161
Diameter, 5

Dimension, 23
Distance, to a subspace, 56

E
Eigenspace, 124
Eigenvalue, 123
Eigenvector, 124
Embedding
    canonical, 93
    continuous, 41
Equation
    homogeneous, 121
    ill-posed, 51
    inhomogeneous, 121
Expansion, binomial, 136
Extension
    continuous, 40
    linear, 71

F
Fredholm alternative, 121
Functional
    continuous linear, 63
    Minkowski, 78
    norming, 75
    sublinear, 71

G
Gram–Schmidt process, 146
Graph, 35

© Springer Nature Switzerland AG 2020
C. Clason, *Introduction to Functional Analysis*, Compact Textbooks
in Mathematics, https://doi.org/10.1007/978-3-030-52784-6

H
Half-space, 78
Homogeneity, of a norm, 19
Hull
    convex, 108P
    linear, 25
Hyperplane, 78

I
Identity, 37
    compact perturbation of the, 119
    parallelogram, 137
    Parseval's, 144
    polarization, 137
Index, 120
Inequality
    Bessel's, 144
    Cauchy–Schwarz, 136
    Hölder, 64
    Minkowski, 28
    triangle
        for a metric, 3
        for a norm, 19
Interior, 5
    algebraic, 46
    topological, 5
Isometry, 41
Isomorphism, 41
    Riesz, 150

K
Kernel, 35

L
Lemma
    core–int, 46
    Ehrling's, 118P
    Riesz's, 24
    Schur's, 100N
    Zorn's, 72
Limit, 5

M
Mapping
    bounded, 6
    closed, 52
    continuous, 6
        sequentially, 7
    open, 48
    quotient, 59
    sesquilinear, 135

Method
    direct, 106
    of successive approximation, 53P
Metric, 3
    discrete, 4
    Euclidean, 4
    product, 4
    relative, 4
    supremum, 16P

N
Neighborhood, 4
Nondegeneracy
    of a metric, 3
    of a norm, 19
Norm, 19
    p-, 28
    equivalent, 19
    Euclidean, 23
    operator, 37
    product, 51
    quotient, 56
    spectral, 37
    supremum, 26, 30

O
Operator
    adjoint, 83
        Hilbert-space, 152
    antilinear, 150
    biadjoint, 94
    bijective, 40
    bounded, 36
    compact, 111
    continuous, 35
        completely, 112
    derivative, 38
    Fredholm, 120
    Fredholm integral, 118P
    injective, 40
    integration, 115
    inverse, 40
    invertible, 40
        continuously, 40
    left-shift, 84
    linear, 35
    multiplication, 131P
    normal, 157
    restriction, 90P
    right-shift, 84

self-adjoint, 157
surjective, 40

P
Pairing, duality, 63
Point
    cluster, 6
    interior, 5
Preimage, 7
Principle, uniform boundedness, 47
Problem, functional, 81P
Product, inner, 135
Projection
    canonical, 51
    metric, 139
    orthogonal, 141

Q
Quadrature, 52P

R
Radius, spectral, 127
Range, 35
Representative, Riesz, 149
Resolvent, 124

S
Semicontinuity
    weak lower, 101
    weak-∗ lower, 101
Seminorm, 55
Sequence
    Cauchy, 6
    minimizing, 106
Series
    convergent, 22
        absolutely, 22
    Fourier, 146
    Neumann, 124
Set
    bounded, 5
    closed, 4
        weakly, 102
    compact, 9
        open cover, 9
    convex, 46
    dense, 5
    equicontinuous, 15
    open, 4
    precompact, 9
    relatively compact, 13

resolvent, 124
sequentially compact, 9
    weak, 103
    weak-∗, 103
totally bounded, 9
Space
    Banach, 20
    bidual, 93
    compact, 9
    complete, 6
    dual, 63
    equivalent, 5
    Hilbert, 136
    inner product, 135
    isometrically isomorphic, 41
    isomorphic, 41
    locally convex, 73N
    metric, 3
    normed vector, 19
    null, 35
    pre-Hilbert, 135
    quotient, 55
    reflexive, 94
    separable, 5
    topological, 7
Span, 25
Spectrum, 123
    continuous, 123
    point, 123
    residual, 123
Subsequence, 6
Subspace, 21
    invariant, 124
Supremum, essential, 31
System
    orthogonal, 143
    orthonormal, 143

T
Theorem
    Arzelà–Ascoli, 15
    Baire category, 45N
    Baire's, 45
    Banach–Alaoglu, 103
    Banach–Nečas–Babuška, 91P
    Banach–Steinhaus, 47
    Bolzano–Weierstraß, 13
    bounded inverse, 50
    closed graph, 51
    closed range, 88

Eberlein–Šmulian, 105
Fischer–Riesz, 146
Fréchet–Riesz, 149
Hahn–Banach, 71
  extension, 74
  separation, 78
  separation, of sets, 80
  separation, strict, 80
Heine–Borel, 12
Hellinger–Toeplitz, 163P
Lax–Milgram, 142
Mazur's, 108P
open mapping, 49
projection, 139
Pythagorean, 138
Radon–Riesz, 67N

Riesz representation, 149
Schauder's, 116
spectral, 161
  little, 129
  reverse, 163P
spectral polynomial, 128
Tonelli's, 106
Weierstraß, 14
Topology, 5, 7
  metrizable, 7
Triple, Gelfand, 154

V
Vector
  orthogonal, 138
  unit, 65
Vector space, *see* Space